Sir Ranulph Fiennes is a man steeped in military tradition, whose career has been spent exploring the world's harshest region. His father served with distinction with the Royal Scots Greys in north Africa, before dying of wounds received fighting in Italy in the Second World War. Ran would later join the same armoured regiment, as a tribute the father he never knew. Armed with an insatiable appetite for adventure, Fiennes was removed from the SAS for misuse of explosives, but, after joining the Sultan of Oman's Armed Forces, he received the Sultan's Bravery Medal on active service in 1971.

Fiennes was the first man to reach both poles by surface travel, and the first to cross the Antarctic continent unsupported. He is the only person to have been awarded two clasps to the Polar Medal for both the Antarctic and Arctic regions. He has led over thirty expeditions, including the first polar circumnavigation of the earth. In 2003, he ran seven marathons in seven days on seven continents, in aid of the British Heart Foundation.

In 1993, Her Majesty the Queen awarded the Order of the British Empire (OBE) to Fiennes, who, on his way to breaking records, has also raised over £14 million for charity. In 2009, he became the oldest Briton to reach the summit of Everest.

THE ELITE

*The Story of Special Forces –
From Ancient Sparta to the War on Terror*

RANULPH FIENNES

SIMON &
SCHUSTER

London · New York · Sydney · Toronto · New Delhi

First published in Great Britain by Simon & Schuster UK Ltd, 2019
This edition published in Great Britain by Simon & Schuster UK Ltd, 2020

3 5 7 9 10 8 6 4 2

Simon & Schuster UK Ltd
1st Floor
222 Gray's Inn Road
London WC1X 8HB

www.simonandschuster.co.uk
www.simonandschuster.com.au
www.simonandschuster.co.in

Simon & Schuster Australia, Sydney
Simon & Schuster India, New Delhi

A CIP catalogue record for this book is available from the British Library

Paperback ISBN: 978-1-4711-5663-2
eBook ISBN: 978-1-4711-5664-9

Typeset in Sabon by M Rules
Printed in the UK by CPI Group (UK) Ltd, Croydon, CR0 4YY

For Charlie Burton –
brave and adventurous and a great companion in adversity.

CONTENTS

INTRODUCTION

Stories involving incredible courage, in the face of unfavourable odds, have always inspired me. From my schooldays at Eton, I would sit enthralled, listening to the adventures of the likes of Scott, Shackleton and Hillary, all venturing to far-off lands, attempting feats most thought to be impossible. Hearing such thrilling escapades eventually drove me to embark on my own adventures around the globe. I'm also quite sure that their examples of bravery have helped drag me out of many a hairy encounter along the way. However, no story has captivated me more than that of 'Colonel Lugs' and the Royal Scots Greys. For Colonel Lugs was my father, and, while we were never to meet, I seem to have spent my life trying to emulate him, and it is his story that has inspired me to write this book.

My father joined the Royal Scots Greys when he was just eighteen. At the time, the Greys were a cavalry regiment, who were famous for using grey horses in battle, most notably in the Battle of Waterloo. On that muddy Belgian battlefield, just as Wellington's British forces looked to be overwhelmed by Napoleon's infantry, the horses of the Scots Greys thundered into the French line. With wild abandon, they viciously bayoneted one French soldier after another, even striking down their drummers and fifers without mercy. On seeing this dramatic change in fortunes, the previously demoralised 92nd Highlanders yelled 'Scotland for ever!' and, clad in kilts, roared into battle. This juggernaut of aggression forced the French back, giving the allies vital time to regroup, and eventually win the Napoleonic Wars in the most dramatic reversal of fortunes. For their heroic deeds, the Greys suffered particularly heavy losses but their sacrifice, just as the war had seemed lost, was for ever embroidered into military history.

Such stories of courage saw the Greys exalted and their beautiful grey horses known far and wide. It is for this reason my father was determined to join them. However, by the time the Second World War erupted in 1939, the regiment was not only about to undergo significant change, but also to face its greatest test, in which my father would play a crucial role.

On the outbreak of war, the Greys were initially sent to Palestine, to help keep the peace between the Jews and the Arabs. It seemed that their days as an elite fighting force were over. During the First World War, in the face of trench warfare and the machine gun, their horses had been virtually redundant, while the rapid advance of the tank and aircraft in the interim had also changed the military landscape. Horses now seemed a remnant of the past on the modern battlefield.

However, as the Greys were being put out to pasture in Palestine the war was turning against the British in north Africa. With the Italians in retreat, Rommel's panzers had come to the rescue. Advancing on all fronts, they had destroyed Britain's under-gunned, undermanned and under-armoured tanks, and looked to be an unstoppable force.

With British forces floundering and facing defeat, the decision was made to redeploy the Greys in more ways than one. Firstly, my father and his men would be sent to north Africa. Secondly, with their grey horses of little use in such a battle, the fateful decision was made to retrain the men for tank warfare. The prime minister, Winston Churchill, an ex-cavalry officer himself, said of this,

'It has been heart-breaking for me to watch these splendid units waste away for a whole year ... These historic regiments have a right to play a man's part in the war.'

The task ahead of the Greys was monumental. Not only would my father and his men have to retrain and somehow master these mechanical beasts, they would then have to face off against Rommel's fearsome division, who had been sweeping all before them. Psychologically, this must have been tough. A big reason many had joined the Greys was to experience the thrill of riding a horse into battle in the open air. Now they had to say goodbye to their beloved animals, of whom they had grown very fond.

From feeling the wind against their faces, the men now had to endure being crammed into a small, hot, metal box. With the hatch battened down, to prevent a grenade being lobbed inside, the conditions in the desert heat were unbearable. At times, with barely any fresh air available, it felt like being baked inside an oven. Because of this, like many others, my father ended up dispensing of his uniform and wore only shorts and canvas shoes. The smell inside was also unpleasant, not so much from the men's sweat, but because the ventilator fans struggled to extract the cordite fumes after the main armament was fired. In time, the Greys would grow used to these conditions but how they must have yearned for their horses in those early days.

The training was also relentless, with the men bruised black and blue at the end of most days. Trundling over rough and uneven terrain, often for hours on end, the crew would bang their joints up against the protruding pieces of hot metal. There were certainly no home comforts either. Resources were so tight that the training tanks had no intercom, so instructors had to tie string to the trainee driver's arms and guide them accordingly. Despite all of these obstacles, my father and his men did not complain. There was a war to be won, and, if they were to overcome Rommel's panzers, they knew they had to focus all their energy on the job at hand.

Yet, after just a few short months of training, the British military could wait no more. Whether or not the Greys felt they were ready for action was unimportant. If they didn't meet Rommel's panzers in battle soon, then all would be lost. Thankfully, the British had upgraded their inferior tanks in the interim to the far superior American Grant tank. At least the Greys would have a machine whose armour and guns could finally match the Germans', even if they were still novices on the battlefield.

Soon after, the Greys were hurled onto 1,400 miles of stony, scrub-covered desert, stretching between Cairo and Tripoli. Against all the odds, with shells bursting, tracers flying and minefields a constant hazard, they somehow held Rommel's panzers at Alam el Halfa, then helped to turn the tide of the war by overcoming them during the Battle of El Alamein.

Over the course of those three days, the men suffered through the most appalling conditions. Cooped up inside their tanks, with no time to sleep, they manoeuvred their way through minefields and rain-sodden mud. All the while, they did battle with the ferocious German tanks and desperately tried to avoid the bombs being dropped from above by Italian planes. During this, my father shrugged off repeated injuries, the worst of which saw shell fragments just missing his femoral artery and what he described in a letter home to my mother as 'other vital bits'. But the Greys' training, firepower and refusal to yield soon saw Rommel's panzers forced into a humiliating retreat. Just four months previously, the German tanks had crushed the Allies in Libya, where they had captured 25,000 troops. Now they were on the run.

Churchill later described this battle as the 'turning of the Hinge of Fate'. He wrote, 'Before Alamein, we never had a victory: after Alamein, we never had a defeat.'

While I beam with pride at the thought of my father leading his men to such a heroic victory, I believe the true measure of the man can be seen in the aftermath, as described by Lieutenant Colonel Aidan Sprot, in his fantastic book *Swifter Than Eagles*:

> Colonel Lugs, during one of his recces, saw two Italians lying wounded away out in No Man's Land. He took Astra out to pick them up while all available enemy arms were directed at him. He and Alec were lucky in that the only damage was a 20mm through a bogey wheel and a 5mm explosive bullet which hit the A/A gun and spattered their faces.

While my father had courageously and gallantly led his men into battle, it is stories such as this – going to the aid of the wounded, no matter for whom they fought, putting his own life on the line in the process – that made me idolise him. However, despite these momentous victories in north Africa, the war continued to rage throughout Europe. My father barely had time to enjoy the moment before he received orders to prepare for another mission impossible: the invasion of Italy.

As with the later D-Day landings in France, the Allies achieved a partial surprise by landing on the Salerno beaches against moderate German defences. The Greys, and their new Sherman II tanks, led the advance for some 2 miles inland before the enemy regrouped and forced an infantry retreat back to the beaches. I am told that, upon this retreat, my father stood in the middle of the road and, after being asked by a group of fleeing soldiers, 'Which way to the beaches?', he replied in his calm, unhurried manner, 'You're going the wrong way, the beaches are over there,' pointing towards the front line. The soldiers sheepishly turned back, and helped to keep the Germans at bay.

At this time, it seems my father encountered a difficulty that I have also faced all my life – how to pronounce our surname. Sprot writes:

Two troopers, who were taken prisoner, escaped back and told us that, when they were being questioned, the German officer said to them, 'We know your squadron leaders are Borwick, Roborough and Stewart, and your commanding officer is F-I-E-N-N-E-S, but we don't know how to pronounce it. Is it Fee-ens or Fines?'

For the avoidance of doubt, I can declare once and for all, on behalf of myself and my father, that the correct pronunciation is 'Fines'. In 2011, the *New Yorker* magazine even ran an article pointing this out in relation to my thespian cousin, Ralph. It seems we Fienneses have always had our work cut out in this regard.

In any event, just as it looked as though the Germans would force the British back into the sea, the Greys unleashed a devastating counterattack. Striking across the enemy's flanks, they restored their position, then chased the retreating Germans up the toe of Italy until the whole of the Neapolitan plain lay below.

Just days after a humiliating defeat had looked on the cards, the Greys entered Naples to the cheers of grateful crowds. But, even in this moment of celebration, my father had no time to revel in glory. Sprot recalls, 'While we were forcing our way through the melee, one would see the colonel charging about with a pick-helve knocking civilians off the tanks!'

General McCreery, the overall Allied commander, later thanked the Greys for the vital part they had played at Salerno. He claimed that it was entirely due to them that the Allies were not thrown off the beaches. Another tribute later came from Major General Graham, the officer in charge of the 56th Division at the landings, who wrote:

At the Salerno landing, I was indeed fortunate to have to work with my Division such a grand Regiment as the Greys. I shall never forget all they did at the time. There were some anxious moments but all was well in the end. That it was so was largely due to the steadfastness and indomitable spirit of your Regiment. There are many glorious episodes in your history, but what you did at Salerno will bear comparison with any.

Following this success, my father returned home to be with my mother for a few days, and it was during this time I was conceived. Yet he was never to know. Just weeks later, he returned to the regiment and, as he set off in his Dingo Jeep to check out enemy positions, he drove over a mine. While he did not immediately succumb to his wounds, he was to perish days later in hospital. Sprot writes warmly of my father on his passing: 'This was a sad day, for the colonel was loved, admired and respected by all in the Regiment, and no finer officer had ever been in it.'

From all that I have read, and been told, my father was not only a legendary commander, but he was also a fine man. It is with immense regret that we were never to meet. As I grew older, and heard of his exploits, I saw it as my destiny that I too would one day follow in his footsteps, seeing action with the Greys in far-flung deserts and jungles, and serving my country by fighting the enemy of the day, which back in the time of the Cold War was the Soviet Union.

With this ambition driving me on, I eventually attended the Mons Officer Cadet School at Aldershot and soon after managed to pass into the Royal Scots Greys as a second lieutenant. For three years, I served as a troop leader, with twelve men and three 70-ton Conqueror tanks at my disposal. With a taste for danger, and an ambition to serve in the most elite units the British Army could offer, I soon set my eyes on joining the SAS. However, after completing the training course I was sadly thrown out in ignominy (more on which later). Undeterred, I volunteered to serve in the Sultan of Oman's Armed Forces in Dhofar for two years, during a time when the communist armies of China and the Soviet Union were looking to invade.

As a consequence of all of the above – my father's courageous deeds, coupled with my own experiences in elite fighting units, as well as an avid interest in stories that involve impossible odds and incredible bravery – I have been inspired to write this book.

With great relish, I have cast my eyes back over 5,000 years of history, revelling in the tales of some of the most remarkable military units of all time. Whether it be fighting on the battle-field, storming forts and castles, rescuing hostages, high-stakes reconnaissance missions or the dramatic assassination of enemy leaders, these men have frequently succeeded against all odds and have often changed the face of history in doing so. But what has been the key to their successes? And what ultimately led to their downfalls? I have sought to answer these questions, and more, while also putting some forgotten, yet heroic, figures into the spotlight, including my father.

1

THE IMMORTALS

On the sandy banks of ancient Babylon's River Tigris, Cyrus the Great's massive Persian army gathered outside the fortified city of Opis. This was a place of considerable strategic importance for Cyrus. If he could take Opis, then Babylon would inevitably fall, just as so many countries had done before it. But Opis was well defended by the Babylonian king Nabonidus, and Cyrus had been unable to break the siege through military might alone. To all intents and purposes, there appeared to be no way in. Yet, ever since his birth, Cyrus had defied the odds. Indeed, it is a miracle that he had even survived to this point.

The Greek historian, Herodotus, claimed that Cyrus's grand-father, the Median king, Astyages, had a dream that signified that his as yet unborn grandson was a threat to his rule. Not wishing to take any chances, Astyages subsequently ordered one of his most trusted men, Harpagus, to kill the infant child upon his birth.

Unsurprisingly, this was not a task Harpagus particularly wanted to undertake. He subsequently passed it on to Mithradates, one of the royal shepherds. However, Mithradates also did not wish to murder a baby, particularly as his own wife was still grieving after having given birth to a stillborn child. At this, Mithradates concocted a plan. Taking the body of his stillborn child to the royal constabulary, he claimed it to be that of Cyrus. The constabulary had no reason not to believe him and took him at his word. With this, Mithradates proceeded to raise Cyrus as his own.

Despite the need for great secrecy regarding Cyrus's true identity, the young boy still enjoyed playing the role of king with his playmates. In fact, he played the role a little too well. During one game, he so severely punished the son of a respected man of the Median community that Cyrus was called before King Astyages to be punished. However, upon seeing the 10-year-old Cyrus, the king immediately realised that it was none other than his own grandson. Harpagus had evidently not carried out the task as had been ordered. The punishment for such duplicity was severe. Herodotus records that Astyages proceeded to invite Harpagus to a banquet, where he cruelly served him the flesh of his own son.

Despite this, Astyages decided to spare Cyrus, sending him to Persia, where he rejoined his family. It was a decision he would soon regret, as it seems that his premonition had been true. In 553 BC, Cyrus rebelled against Astyages, with the vengeful Harpagus by his side. Going on to conquer the Median Empire, Cyrus mercifully spared his grandfather, but kept him in his court until his death. No doubt Astyages rued the day he had not ensured his grandson's death.

After establishing himself as the new ruler of the combined Persian and Median kingdoms, Cyrus rapidly expanded his empire, whether through marriage, negotiation, outright slaughter or ingenuity. For instance, when meeting the Lydian forces in 546 BC Cyrus created an improvised camel corps from his baggage train, and placed the baying animals at the front of his formation. The unaccustomed smell and sight of the camels was reported to have thrown the Lydian cavalry into disarray and helped lead Cyrus's forces to victory.

Such tactical brilliance soon saw the Persian Empire cover an area that included the modern countries of Turkey, Armenia, Azerbaijan, Iran, Kyrgyzstan and Afghanistan. The only remaining significant unconquered power in the Near East was the Neo-Babylonian Empire, which controlled Mesopotamia, as well as kingdoms such as Syria, Judea, Phoenicia and parts of Arabia. And Cyrus aimed to take them all, with the help of his elite special forces: the 'Immortals'.

Unfortunately, beyond the writings of Herodotus, and Persian paintings or sculptures, historical knowledge of the Immortals is somewhat limited and inconsistent. Herodotus said of them:

> Of these 10,000 chosen Persians the general was Hydarnes the son of Hydarnes; and these Persians were called 'Immortals', because, if any one of them made the number incomplete, being overcome either by death or disease, another man was chosen to his place, and they were never either more or fewer than 10,000.

While there are few specifics as to their training or recruitment, Herodotus claims that, from the age of five, all Persian boys would be 'taught to ride, to use the bow and to speak the truth'. Such skills were practised and honed until the age of twenty, with the most important being the ability to tame a wild horse, as the Persians were renowned for their impressive horse-riding ability.

Following this, the Greek philosopher Strabo tells us that compulsory military service lasted from the age of twenty to twenty-four, where the young men were divided into companies of fifty, each under the command of a son of the nobility. After service, Strabo claims that the young men were demobilised but remained liable for military service until their fiftieth year. However, the very best of these men were recruited to the ranks of Cyrus's elite killing machine.

With such incomplete information available, there are conflicting theories about how the Immortals lined up, and about what actual function they served. Many believe the Immortals were primarily infantry, though there may have been some cavalry. Some believe there was even an elite within an elite, whereby 1,000 of the Immortals' best men served as bodyguards to the king, who were signified by having apple counterbalances on their spears. Pictures and sculptures from the time certainly prove that they carried spears, with the regular 9,000 having silver or golden pomegranate counterweights. They were certainly well looked after, travelling to battle in luxury caravans, with servants attending to their every need. The finest food and drink were always available, as was a harem of beautiful women. Whatever their purpose, it was the Immortals' unique set of skills that helped Cyrus conquer kingdom after kingdom, with Babylon next on his agenda.

For months, Nabonidus had kept Cyrus and his Immortals at bay. Three walls of the Opis fortress were well defended by his soldiers, while the fourth backed onto the Tigris River. But this was a weakness Cyrus sought to exploit. If he could somehow get rid of the water, then he could attack the wall where the Babylonian defences were at their weakest.

As such, Cyrus's engineers had spent weeks working to divert the water. Now the river outside the fortress was dry, and the path to Opis, and all of Babylon beyond it, lay open for Cyrus. All that stood in his way were Nabonidus's forces and, as with so many before them, Cyrus and the Immortals were confident they would again overcome and add yet another kingdom to the ever-expanding Persian Empire.

In the blazing sun, the 10,000 Immortals waited for their king to give his order to attack. Surrounding him were his 1,000 elite guards, with apple counterbalances on the end of their 6ft spears. Whether these so-called apple bearers were officially 'Immortals' it is hard to say, but the king trusted them with his life. In front of them stood the main bulk of the Immortal forces, the sun shimmering off their bronze breastplates and helmets. To protect their faces from the dust being kicked up by the desert breeze, or the horses' pounding, some of them wore a cloth cap, known as a Persian tiara, which they pulled over their faces. In their hands, they held their 6ft spears, some with golden pomegranate counterbalances, others with silver, their knuckles turning white in anticipation of the battle ahead. Meanwhile, the Persian archers stood before them, either side of the cavalry, their bows drawn and ready.

Suddenly, the call to attack echoed across the line. The archers immediately unleashed a torrent of arrows that blotted out the sun and on impact pierced the flesh of thousands of Nabonidus's men. As screams filled the air, the cavalry thundered forward. With slingshots in their hands, they fired stones with unerring accuracy, knocking the enemy to the ground, taking out their eyes and fracturing their skulls with a vicious crack. The Immortals now followed. Some launched their spears over their heads, sending them soaring through the air and plunging into the chests of the enemy. Others ran into battle, raising their shields for protection in one hand, stabbing with their spears in the other, moving relentlessly onwards. Should they be outnumbered, they stabbed in one direction then used their heavy counterbalances to deliver a knockout blow in the other. With some spears breaking in two on impact, they reverted to using their daggers, slings, maces or bow and arrow to maim and kill any man who stood in their way.

As always, this form of attack was brutally effective. Before long, hardly any of the enemy were left standing, while any casualties or injuries in the Immortals' ranks were immediately replaced by fresh recruits.

Surveying the dead bodies littered across the battlefield, Cyrus looked with satisfaction across the horizon – Opis was his, and the road to Babylon was now clear. Just days later, he seized the city and proclaimed himself 'king of Babylon, king of Sumer and Akkad, king of the four corners of the world'.

However, while Cyrus became loved and adored in Babylon, and went on to expand his empire with the help of his Immortals, he was to meet a grisly death in 530 BC. There are many differing accounts regarding the precise nature of Cyrus's death, but Herodotus claims that, after a savage and lengthy campaign against the Massagetae tribe, their queen, Tomyris, cut off Cyrus's head and plunged it into a barrel of blood, apparently to signify his supposed insatiable appetite for blood and conquest.

Yet, even as Cyrus perished and was succeeded by his son Cambyses II, the Immortals endured. While they continued to guard their king and help expand his empire, they soon faced a situation that military might alone would not allow them to overcome.

Although the early years of Cambyses' reign were relatively uneventful, there were already signs of things to come when he killed his young brother Smerdis, after seeing him as a rival. Herodotus has said of him, 'They say that from birth Cambyses suffered from a serious illness, which some call the sacred sickness,' and, 'I have no doubt at all that Cambyses was completely out of his mind.' Indeed, it was his rage at the broken promise of marriage that saw Cambyses embark on his first major conquest.

Herodotus claims that this situation arose when the Egyptian Pharaoh Amasis II decided against marrying off his own daughter to Cambyses, despite a promise to the contrary. He instead sent the daughter of the former pharaoh, Apries, in her place. Insulted and aggrieved, when the poor girl arrived at Cambyses II's court, she immediately revealed Amasis' deceit.

Cambyses was outraged and vowed revenge for such an affront. However, just as he mobilised his troops Amasis suddenly died, leaving Egypt in the hands of his son, Psametik III. But this did not deter Cambyses. By now, his heart was set on conquering Egypt and the death of Amasis would only make this easier than he had envisaged. After all, Psametik III was just a young man who had lived largely in the shadow of his father. And the odds against him were getting bigger with every passing day. Upon hearing of the Persians' planned invasion, Psametik's Greek allies had abandoned him. Meanwhile, his military counsel, Phanes of Halicarnassus, anticipating the slaughter to come, betrayed him and actually joined the Persians' side, becoming a key source of intelligence.

With little option, Psametik decided to fall back on the tried and tested method of defence in such times: he fortified his position at Pelusium, near the mouth of the Nile. Awaiting the Persian attack, he hoped he could frustrate them long enough to find a better solution. In such siege situations, the Immortals were by now well aware that brute force rarely sufficed. If they were to attempt to storm the walls, they would be met by a hail of arrows, spears and boiling oil. Despite this, Cambyses was impatient to end the siege quickly and enjoy the spoils of victory.

While there are differing versions of what occurred next, I particularly like that of the second-century AD writer, Polyaenus. In his *Strategems*, he recounts that, while the Egyptians were initially successful in holding back the Persian advance, it was Cambyses' knowledge of Egyptian beliefs and traditions that ultimately held the key to success.

In ancient Egypt, cats were not only a popular pet but they were also revered for being closely associated with the goddess Bastet, who appeared in Egyptian art with the body of a woman and the head of a cat, or as a sitting cat in a regal pose. If she should be insulted, it was said that she would inflict plague and disasters on humanity. One way to ensure insulting Bastet was to harm a cat. Herodotus has stated that cats were so highly regarded in ancient Egypt that the punishment for killing one was death. There are also stories in which Egyptians caught in a burning building preferred to save their cats before saving themselves. (On a side note, as an avid lover of cats, I believe this is exactly what my wife, Louise, would do in such a situation. Furthermore, in houses where a cat has died a natural death, its inhabitants would shave their eyebrows as a sign of their grief. This is definitely a tradition I will not be sharing with my wife.)

It was this knowledge of Egyptian culture and values that suddenly gave Cambyses a brilliant idea. He quickly ordered the Immortals to paint the image of Bastet on their shields and then round up as many of the animals as possible. It is quite amusing to think of these elite warriors scouring the countryside for these wild animals, but their efforts would prove to be worth it.

On Cambyses' orders, the Immortals moved forward on Pelusium, in front of a line of thousands of meowing and screeching cats. The Egyptians were befuddled. They daren't shoot at the animals, or risk marking Bastet's image on the Immortals' shields, for fear of insulting the goddess. As they dithered, the Immortals moved ever closer, led by their army of cats. As I write this, I can see my wife's fourteen cats (yes, you read that correctly, fourteen!) all patrolling the house and the gardens. The thought of fourteen ganging up on me is not a pleasant one, so imagining thousands of these creatures going into battle is enough to make me sympathise with what happened next.

Unbelievably, as the Immortals approached the fortress, the Egyptians surrendered their position and took flight. At this sign of weakness, the Immortals showed no mercy, viciously striking down all who tried to escape. The slaughter was so great that Herodotus reported seeing the bones of the Egyptians still in the sand many years later. And, while those Egyptians not killed at Pelusium fled to the safety of Memphis, the Persian army was in full pursuit and soon took Memphis too, with equally bloody consequences, as Psametik was captured and executed.

With this act, Cambyses and the Immortals ended the sovereignty of Egypt and annexed it to Persia. However, such was Cambyses' scorn for the Egyptians' beliefs, and their pathetic defence, that, after the battle, he hurled cats into the faces of his defeated enemy, incredulous that they would surrender their country, and their freedom, for these common animals.

From this behaviour, it will not surprise you to learn that Cambyses soon came to resemble the stereotypical mad king. While he defiled and burnt the corpse of the pharaoh Amasis, he proceeded to slay a newborn Apis calf, worshipped as a god in Egyptian religion. Yet it seems the very beliefs he mocked would eventually have their revenge. Herodotus claims that, in April 522 BC, Cambyses died after his sword slid out of its scabbard and pierced his thigh, in the exact same place where he had stabbed the sacred Apis calf years before.

While Cambyses' successors ensured that the might of the Persian Empire continued, with more countries and continents falling under their command, one remained elusive: Greece.

When Xerxes I took the Persian throne in 486 BC, Greece became his obsession. However, it would take more than cats for Xerxes to bend the country to his will. For, while only 300 men stood in the way of his massive army, these men were regarded as one of the greatest elite military units the world has ever known ...

2

THE SPARTANS

480 BC

As the fierce sun beat down over the Spartan lands of ancient Greece, hundreds of thousands of Persian infantrymen marched to the narrow shore of Thermopylae. After taking the Greek capital of Athens, the Persian emperor, Xerxes, now looked to complete his conquest. And, for him, the conquest of Greece was personal, owing to his father's failed attempt to claim it a decade earlier at the Battle of Marathon. Legend has it that, following this Greek victory, the official Athenian messenger ran all the way (26 miles) back to Athens to announce the Persians' defeat. This was the origin of the marathon run, something I know through gruelling experience, having run seven marathons, on seven different continents, in seven days in 2003, just a few months after suffering a heart attack. Thankfully, I survived such a bitter ordeal, while the poor Athenian messenger did not.

In any event, the Persian invasion of Greece had been years in the making. While plans were perfected, intensive conscription had swollen the hordes of the Persian army to almost a million men. A relentless force, all that now seemingly stood in the way of Xerxes' supposed destiny was a group of 300 men, with long oiled hair and thick beards.

Guarding a narrow passageway between the ocean and the mountains, these men blocked the only route to the southern regions of Lokris and Thessaly. Yet their presence seemed inconsequential. Countries such as Egypt and India had already bowed before Xerxes and these 300 men, armed with just wooden shields and spears, would surely yield. But the grizzled faces that peered from beneath their Corinthian helmets made it clear that, no matter the odds, they were prepared to die for their cause. Demaratus, the former Eurypontid king, would later say of the Spartans in such situations, 'They do what the law commands and its command is always the same, not to flee in battle whatever the number of the enemy, but to stand and win, or die.'

Almost taking pity on this supposed last stand, Xerxes offered mercy. Sending an emissary to Leonidas, the Spartans' inspirational 60-year-old leader, Xerxes invited them to surrender by handing over their arms. Leonidas's subsequent response has gone down in legend: 'Come and take them.'

This was an invitation Xerxes did not intend to decline. However, not wishing to waste the efforts of his Immortals on such infidels, he duly sent 10,000 of his second rank to finish them off. But, as the first line of Persian troops charged towards the Spartans, Leonidas suddenly ordered his men to group together. What seconds before was 300 disparate men now formed into a solid phalanx, creating a wall of shields and spears, and, with it, a ferocious fighting machine.

Before the Persians knew what they were up against, the first line of men felt the sharp tip of the Spartan spears rip through their flesh. Up close, they then saw the unrelenting black eyes of their enemy, as they took their last breath. The Spartan spear was thrust into their body then removed and planted into the next advancing Persian, the same motion repeated again and again. And if a spear should be broken or jammed into the flesh of the enemy, a sharp sword was always close at hand, eager for the next victim. Yet the Spartans felt no remorse. As one of the world's first elite band of fighters, they had been trained to kill since birth.

A rugged mountainous region of Laconia, Sparta was world renowned for breeding fighters of startling obedience, resilience, intelligence and bloodthirsty ferocity. It was the only Greek state with a full-time professional army and, as such, from the moment a male took his first breath, he was trained for war. This is no exaggeration. Any male babies seen to be weak, or somehow unsuitable for the vigorous training ahead, were dropped into a pit and left for dead. The phrase 'only the strongest survive' has never been more apt. However, it also makes a mockery of all those slow starters in life, who go on to excel in their later years, myself included. When I was young, my mother used to put long blue ribbons in my curly hair, which led to some mistaking me for a girl. In Sparta, I surely would have been destined for the pit!

Sentimentality was one emotion the Spartans clearly lacked. From the age of five, the young Spartan boys were thrown head-long into their military training. Sent away to live in barracks, any sign of weakness was instantly leapt upon and corrected. While they were trained to fight, they also had to walk everywhere barefoot, so as to harden the soles of their feet on the hard terrain, as well as to harden their resolve. The goal was to create obedient, resilient, hardened warriors, and often these young boys would go through training that would even make the men of today's elite special forces wince.

Once a year, the boys were taken to the Temple of Diana, where they were flogged one by one. During this flogging, they were expected to remain silent. If they cried out, this would be seen as a great source of shame. I know how they felt. In the 1950s, when I attended Eton, whenever I was beaten by the senior house prefect for a misdemeanour, I tried hard to keep my mouth shut for fear of being called a wimp.

A grisly story, which emphasises just how seriously the Spartan boys took this, involves a young boy who smuggled a fox cub into school. Hiding it under his cloak during his first lesson, the fox's desperate attempts to escape saw it gnaw and claw at the boy's chest and stomach, cutting him to pieces. Yet the boy never made a sound. Before the lesson ended, he had collapsed and died, but, in the eyes of the Spartans, his honour was intact.

To oversee their progress, each boy was allocated an older guardian, who was known as his 'lover'. Homosexuality between the boys and their 'lover' was accepted, but every Spartan was expected to eventually get married. How else would the Spartans provide a constant supply of warriors for their killing factory?

As well as learning how to fight, and how to withstand pain, food was also withheld so as to ensure the boys would become self-sufficient. In order not to starve to death, the boys were meant to forage or steal their food, without getting caught. To encourage this, the Spartans even held an annual festival where chunks of cheese were left on rocks, guarded by men wielding weapons. The boys were expected to approach with stealth and cunning if they were to get the cheese without being detected. And, while there was the threat of a beating if they were caught, there was the double bonus of food, and respect, if they were not.

By the age of eighteen, the boys were conscripted into the Spartan army and issued with their armour and weapons. However, by far their most treasured possession was their shield. Demaratus of Sparta stated that, during combat, a hoplite (a heavily armed foot soldier) could lose his helmet or his weapon but, if he misplaced his shield without a good, proven and witnessed military reason, then 'he be disgraced'.

Legend has it that mothers would tell their sons upon going into battle, 'Come back with this shield or on it.' The only choice any Spartan had was death or victory. Indeed, there was no tolerance for cowardice in Spartan society. When a traitor to Sparta took refuge in a sanctuary, rather than plead for his life, his mother was said to have taken a brick and placed it in the doorway. Following this example, the Spartans bricked up the temple door with the traitor left to die inside. He might have considered himself lucky. Any soldier accused of cowardice could be beaten up with impunity, forced to wear coloured 'coward' patches on his shirt and made to shave off one-half of his beard. Such soldiers were also not allowed to marry; presumably for fear that their offspring would inherit their cowardice.

Yet there was good reason the shield was considered so vital, as any man who lost it when in the phalanx could be putting the whole structure at risk. As Plutarch, the Greek biographer, explained, 'They wear their helmet for their own sakes but carry their shields for the whole line.' Nic Fields, in his book *Thermopylae 480 BC: Last Stand of the 300*, laid out why the shield was so crucial in this regard:

> It was the hoplite shield that made the rigid phalanx formation viable. Half the shield protruded beyond the left-hand side of the hoplite. If the man on the left moved in close, he was protected by the shield overlap, which thus unguarded his covered side. Hence, hoplites stood shoulder to shoulder with their shields locked. Once this formation was broken, however, the advantage of the shield was lost; as Plutarch says (Moralia 241), the body armour of a hoplite may be for the individual's protection, but the hoplite's shield protected the whole phalanx.

For those Spartans who displayed exceptional fighting and survival skills, there was yet another prize – entry into the Spartans' elite special forces, the 300. And, if the 300 was not prize enough, for a chosen few, there was the chance to become a member of the so-called Krypteia, the most elite Spartan military group of all.

As you might expect, entry into the Krypteia was far from easy. According to Plutarch, every autumn the leaders of Sparta would declare war on its helot (servant) population. Any soldier wishing to join the Krypteia could therefore kill a helot without fear of repercussions. In other words, the state sanctioned murder against its own citizens in order to strengthen its elite military unit. At night, the hopefuls would descend on the Laconian countryside, armed with knives, and would hunt down a helot to earn their stripes. But just killing any helot wouldn't do. To really prove their worth, they aimed to kill the strongest in cold blood, all without being caught. While becoming a member of the Krypteia was considered very prestigious in its own right, it was often the only way for a Spartan to rise to the ranks of leadership in Sparta, as their king, Leonidas, had once done.

Leonidas was, however, said to descend from the dynasty of the mythological demigod Heracles. Such lineage meant he was always destined to be king, but there was also no doubt that the young Leonidas was an incredibly capable warrior, and leader, who earned the respect of his men the hard way. Plutarch said of his self-assurance: 'When someone said to him, "Except for being king you are not at all superior to us," Leonidas, son of Anaxandridas and brother of Cleomenes, replied, "But, were I not better than you, I should not be king."'

Such was Leonidas's standing in Sparta, and beyond, that, when the Persian hordes sought to invade Greece, the confederated Greek council looked to him to defend it. Themistocles, an Athenian politician, suggested that, in order to conquer southern Greece, Xerxes and his army would need to travel through the narrow pass of Thermopylae. If Leonidas and his men could somehow block this route, then they might frustrate Xerxes just long enough for the Greeks to reassemble their forces.

While such an operation seemed unlikely, Leonidas already saw it, and his death, as his rightful destiny. Herodotus claims that the Oracle, who was consulted about important decisions in the ancient world, made the following prophecy to the Spartan leader:

> For you, inhabitants of wide-wayed Sparta,
> Either your great and glorious city must be wasted by
>   Persian men,
> Or, if not that, then the bound of Lacedaemon must mourn a
>   dead king, from Heracles' line.
> The might of bulls or lions will not restrain him with oppos-
>   ing strength; for he has the might of Zeus.
> I declare that he will not be restrained until he utterly tears
>   apart one of these.

However, with estimates of the Persian forces ranging from 300,000 to over 2 million, Leonidas only selected those Spartans who had living sons for the operation, knowing that they were almost certain to die. But he was also not without hope, for he knew more than anyone the prowess of his men.

With the battle narrowed to a stretch of land between Mount Kallidromo and the sea, Leonidas knew that the Persians would have no option but to face his men head-on if they wanted to advance south. It would be a tiring, ferocious defence, in temperatures reaching over 40 degrees. Yet the Persians would be forced to do most of the running, with the Spartan phalanx just needing to hold firm.

On the first day of fighting, the Spartans' line didn't break. Repelling wave after wave of attacks, their spears killed one Persian every four seconds. Soon mountains of Persian bodies, amid pools of blood, piled up in front of the Spartan phalanx. This only made conditions even more difficult for the Persians. Forced to climb over the bodies of their comrades, they were unbalanced when they finally came face to face with the sea of spears. Before they could even raise their swords, they helpfully fell onto a spear, doing the Spartans' job for them.

With Xerxes' secondary military unit facing catastrophic losses, he decided it was time to put a quick end to matters. As such, he called upon the 'Immortals'. But even the Persian army's most elite force was no match for the 300. With Spartan spears over 2.5 metres long, compared to the Immortals' 2-metre-long weapons, they merely repeated the fate of the thousands of infantrymen who perished before them. By the end of the first day, Persian casualties were in the thousands, while the Spartan 300 suffered only minor losses.

Facing off against a far superior enemy is actually something I have a little experience of. In the late 1960s, I volunteered to serve in the Sultan of Oman's Armed Forces in Dhofar, which were under threat from invading Marxists. With few resources, five Land Rovers and a maximum of thirty men at my disposal, I was expected to engage with the larger, and better equipped, so-called *Adoo* – 'enemy' in Arabic. Upon arrival, I also found my men looked to be unfit for service. Indeed, when Captain Southward-Heyton introduced me to them he warned, 'Between you and me, it would be suicide to go anywhere near *Adoo* territory with this bunch in their present state.' I soon realised that, if we were to survive, I had to be creative.

Night patrols were especially effective, as we could attack unseen, instilling terror into the enemy, while also making them believe we had a far larger force than was in fact the case. Working at night certainly suited my men and me, as daytime temperatures would frequently reach 50 degrees in the shade, making conditions unbearable. Another trick I learnt was, when under attack, if I were to shout 'attack' or 'retreat' at my men, they were to do the opposite. This allowed us to escape into the night while our superior enemy braced itself for the attack they thought was coming.

In daylight, I also made a great effort to showcase our 'strength'. Near *Adoo* territory, I would set off mortars and allow my best marksmen to engage in shooting practice. While the mortars caused no actual damage, and only steel drums were hurt by gunshots, it still showed our enemy what we were capable of. Such psychological warfare was vital if we were to keep far larger forces at bay.

Intelligence on the enemy was also crucial. A key part of this was employing a local bedu guide, who was an expert in camel tracks and droppings. Sometimes he knew the name of a camel's owner just by examining the shape of the hoof mark left in the dirt. By examining the texture of the animal's droppings, he could even tell where the enemy had last eaten. It was truly incredible, not to mention very helpful in planning our patrols.

However, unlike the Spartans, I must admit that I found killing the enemy in close combat very hard to deal with. The first time this happened, the sad and surprised look on the man's face as I shot him, before he had the chance to do the same to me, lived with me for a very long time. I've said previously that I felt a part of me died with him that day.

After a day of ferocious combat, Leonidas and his men retired to the mountains for the night to regroup. Around a campfire, they relished not only buying their Greek compatriots time, as they had promised, but also surviving another night on earth. Barring a miracle, they had all accepted that they would eventually die. They just had to withstand the Persians for as long as possible.

However, as they rested, Leonidas was alarmed to hear that there was a secret path over the mountains, which stretched beyond the Spartan line. If the Persians were to find this, they could bypass the Spartans and encircle them from behind, leaving the phalanx all but redundant. Fearing such an outcome, Leonidas sent a group of Greek fighters to guard the path. If the Persians were to find and cross it, all would be lost.

Meanwhile, Xerxes admonished his generals in his tent. The first day had been catastrophic. While his infantry had suffered thousands of losses, even his elite Immortals had failed to breach the Spartan wall. But, if he was to reach the southern regions, Xerxes had to somehow defeat this stubborn band of 300 men. Devoid of other ideas, Xerxes decided that the Spartans were bound to tire and it was only a matter of time before they broke. With close to a million men at his disposal, he believed he could afford to sacrifice a few second-class infantrymen to wear the Spartans down.

The following day, Xerxes launched a relentless assault on the Spartans. Minute after minute, hour after hour, waves of Persians crashed into the Spartan phalanx. But still they held firm. The intense heat, and little rest, saw their muscles cramp, and their arms and legs tire from holding up their heavy shield and spears, but as the sun set on the second day, thousands more Persians lay dead.

However, that night, catastrophe struck for Leonidas. A local resident called Ephialtes visited Xerxes with momentous news. Behind the Greek lines was a secret path that could lead to victory. Wasting little time, Xerxes sent his Immortals off into the night, where they made quick work of the Greek defence, and looked to surround the Spartans by daylight.

Having escaped slaughter at the hands of the Immortals, a Greek runner rushed to tell Leonidas that their defence had been breached. It was now clear that the phalanx of the 300 would no longer suffice on an open battlefield where they could be attacked on all sides. At this, Leonidas decided to utilise the skills of the Krypteia, with the head of Xerxes, no less, their target. Striking down the Persian leader would not only destroy the cohesiveness of their army, but also send a signal that no Persian was safe.

Approaching the bustling Persian camp at night, the Krypteia wrapped their blades to ensure they didn't shimmer in the moonlight. Finding Xerxes' tent, in this vast sea of men, was the easy part. Not only was it the grandest in the camp, but it was also surrounded by his caravan of slaves and women. However, getting to the tent itself, and through the thousands of Persians, and guards, undetected, would test them to their limits.

Sadly, their fate is lost to history. All we know for sure is their mission failed, as Xerxes survived the night and, at the break of dawn, ordered his men to finally kill off the Spartans. All that was left for Leonidas, and his men, was to somehow buy more time with one last heroic stand. Knowing death was now inevitable, Leonidas kept the best of his fighters with him, while he allowed the bulk of his army to escape in order to fight another day. This was the mark of a true leader.

At daylight, a Persian force of over 10,000 men charged at the last remnants of Leonidas's forces. Even now, after days of fighting, and knowing death was imminent, the Spartans fought until their spears were shattered and their swords broken. Yet even that didn't stop them, as they then continued to fight with their hands and teeth, heroically killing Xerxes' brothers Abrocomes and Hyperanthes.

At this, Xerxes reached breaking point. With the Spartans scattered, he sent for his archers to finish them off. Plutarch claims that, upon seeing this, one of the Spartan soldiers said to Leonidas, 'Because of the arrows of the barbarians it is impossible to see the sun.' Leonidas replied, 'Won't it be nice, then, if we shall have shade in which to fight them?'

As the thousands of archers pulled back their bows, Leonidas issued a final cry to his Spartans, 'Death or victory!' Remembering what their training had taught them when faced with mass archery, they sprinted at the enemy, thus minimising the time they would be exposed to the deadly shower of arrows. Roaring out their battle cry, which was described by the war poet Aeschylus as 'the noise of the scream of eagles', Leonidas and his men fought until every last one of them was dead.

While the Persians claimed victory, they had suffered over 20,000 casualties. And, despite the death of the 300, the Spartans would ultimately endure. A year later the surviving Spartans reassembled with Greek forces and defeated the Persians at the Battle of Plataea, finally putting a decisive end to the Greco-Persian war, and ensuring the Spartans were renowned as the most brutal, and effective, fighting force in ancient Greece.

However, a rival had taken note of their every move, and now planned to use their skills against them. With this, the Spartans would soon be consigned to the history books ...

3

THE SACRED
BAND OF THEBES

379 BC

The Persians might not have succeeded in conquering Greece but, by 400 BC, the country seemed intent on tearing itself apart. Its three major regions, Sparta, Athens and Thebes, each had a wildly different political outlook, which put them at loggerheads. While Athens was considered the model of democracy, Sparta was at the other end of the spectrum, regarded as the most autocratic military dictatorship in the ancient world. Meanwhile, Thebes was somewhere in the middle. Although its citizens clamoured for Athenian democracy, its ruling elite did all it could to keep hold of power.

This ideological battle was one Sparta kept a close watch on. Should Thebes suddenly follow Athens' lead, then the rule of Sparta's two kings, Agesilaus and Cleombrotus, could also come under threat. Agesilaus refused to counter such a situation. In his sixth decade, he had never lost a battle as leader of the Spartan race and was regarded as one of the most powerful kings in history. To Agesilaus, democracy was a threat that had to be crushed at all costs.

Consequently, Sparta embarked on an aggressive unilateralist policy towards the rest of Greece. As the citizens of Thebes rebelled, Agesilaus eventually marched on the city in 382 BC and proceeded to install a puppet Spartan government, consisting of Archias, Leontiades and Hypates. With this, all dissidents were brutally dealt with, causing many to flee the city rather than be imprisoned or executed.

One such dissident who fled to the safety of Athens was Pelopidas. According to Plutarch, his Greek biographer, Pelopidas came from great wealth but gave most of it away to the poor, preferring to live a simple life and immerse himself in the military. Such deeds made him an incredibly popular figure in Thebes, as did his daring, brave escapades on the battlefield, where he earned a reputation as one of the finest soldiers in the region. It was on the battlefield that he met Epaminondas, a fellow soldier who not only saved his life, but in time would also help him overthrow the Spartans.

Sadly, little is known about Epaminondas. What we do know has been cobbled together from accounts in Plutarch's 'Life of Pelopidas', as well as briefly in the writings of Xenophon and Diodorus Siculus. From these sources, we know that, while Pelopidas fled to Athens, Epaminondas stayed in Thebes, where he had turned his back on the military and was now regarded as one of the region's most admired philosophers. As such, the pro-Spartan leadership did not see him as a threat. This was to prove to be a catastrophic mistake.

While Pelopidas built a small resistance group from Athens, Epaminondas did likewise in Thebes. Joined by the likes of Charon and Gorgidas, they spied on the powers that be and fed information back, all the while planning an assault that would rock Thebes, as well as Sparta. And, by December 379 BC, they were ready to put it into action.

From his grand home in Thebes, Archias, the pro-Spartan Theban leader, hosted a raucous party. Joining him were his colleagues Leontiades and Hypates, as well as other nobles and politicians. Satisfied that all resistance had been wiped out, they had now settled into a life of debauchery, growing fat and drunk, with servants catering to their every whim and desire. Feasting on the finest food and wine, they all eagerly anticipated the evening's main event, as twelve scantily clad hetairai, the most beautiful prostitutes in all of Thebes, suddenly waltzed into the room.

Enraptured, Archias immediately beckoned them to join him, so that he might get a closer look and take his pick. As the ladies of the night sashayed across the room, the drunken powerbrokers lustily watched on, eagerly awaiting their turn.

Yet, when one of the girls came near to Archias, he narrowed his eyes; something wasn't right. Before he could act, a dagger was slashed across his throat. The girl was Pelopidas in disguise. As Archias collapsed into a pool of blood, the other disguised hetairai proceeded to slaughter everyone in the room. Pelopidas and Epaminondas's anti-Spartan uprising had begun. Yet the job was only half done. There was still the small matter of the Spartan garrison based in the town, which would be sure to strike back.

However, as word rippled through the streets that Leontiades, Hypates and Archias were dead, the citizens of Thebes decided to take their chance. Breaking into the city's armouries, they armed themselves and marched on the Spartans. Outnumbered, and facing a furious mob, the Spartans left the city without a fight, making a mockery of their 'death or victory' mantra. For now, Thebes was free of the Spartans. Yet everyone was well aware this was just a fleeting victory. King Agesilaus would surely not take this rebellion lightly. Hell was soon going to pay them a visit, hard and fast. If they were to survive the Spartan military juggernaut, the Thebans would need to be prepared.

No army had beaten Sparta's mighty warriors in open battle for over 100 years. As seen in the previous chapter, their training regime was designed to build the toughest warriors on the planet. By contrast, Thebes didn't even have a full-time army. All it had to offer were reservists who had only received the most basic training. Should they come head to head with the full force of the Spartans, it would be no contest. While Thebes soon found allies in the surrounding region of Boeotia and beyond, Pelopidas, Gorgidas and Epaminondas decided they needed to create a full-time elite fighting unit of their own. They named it the 'Sacred Band' and it was destined to become one of the most unique special forces in military history.

Again, much detail has been lost as to the Sacred Band's origins, but we know they took inspiration from the Spartans. Thus a force of 300 of Thebes's most fearsome warriors was raised to go toe to toe with the Spartan 300 – but there was one crucial difference. Not only were all of the 300 Thebans homosexual, they were 150 couples.

While homosexuality was accepted in ancient Greece, it appears there were good military reasons the unit was to be composed of couples. The Spartan 300 had trained and fought together over many years. This had forged close bonds between the men, and meant they were willing to fight to the death for each other. To build such a relationship between men takes time, something Thebes did not have. Placing lovers in combat together was therefore designed to jump-start the process of building esprit de corps. Plato certainly approved of such a decision, observing that men might actually fight better next to their lovers, writing in his *Symposium*:

> And if there were only some way of contriving that a state or an army should be made up of lovers and their beloved, they would be the very best governors of their own city, abstaining from all dishonour, and emulating one another in honour; and when fighting at each other's side, although a mere handful, they would overcome the world. For what lover would not choose rather to be seen by all mankind than by his beloved,

either when abandoning his post or throwing away his arms? He would be ready to die a thousand deaths rather than endure this. Or who would desert his beloved or fail him in the hour of danger?

Historians disagree about the exact date the *Symposium* was written but many believe it was before the Sacred Band was formed. Therefore, it might have even served as inspiration for the Band's formation.

However, one definite source of inspiration arrived courtesy of Thebes's most celebrated son: Hercules. Despite being married, the famous figure from Greek mythology had shared a passionate homosexual relationship with his charioteer, Iolaus, which saw them become a formidable fighting team. Their extraordinary devotion to each other, and legendary victories, was exactly what Epaminondas and Gorgidas hoped to recreate with the Sacred Band.

In all of my endeavours, I have always recognised the necessity of a strong bond within a team. When I volunteered to serve in Oman, I was put in charge of a platoon who were a mix of Omanis and Baluchis, who absolutely hated each other. When your focus should be on fighting the enemy, you certainly don't need to be concerned about your men fighting among themselves. Only able to speak the most basic Arabic, I was also unable to understand what was being said as words flew between them, hot and bitter. I felt about as effective as a pint of oil in a storm-bound ocean. When they argued about whose turn it was to be on watch, I suggested that we flip coins to decide. They looked at me as though I were mad. Such western games were unheard of. With all my initial attempts to find some harmony between them failing, I made it a top priority to fix this issue, before it dragged the whole platoon under.

Of course, if they were going to have any respect for each other they first had to gain respect for their leader. While brushing up on my Arabic, so I could finally understand what was going on, I also endeavoured to treat each man fairly, to avoid any allegations of favouritism. I also ensured I knew each man's name and background. Such familiarity saw them warm to me, so much so that they eventually christened me *Bachait bin Shemtot bin Samra*, which apparently means 'John, Son of Rags, Son of the Thorntree'. I'm not entirely sure if this was an affectionate name but it seemed they had at least come to respect me.

With this, I could start to train my team to fight as a unit, rather than as individuals. I wanted each man to realise why they, and their supposed enemy, were a vital cog in the overall machine. Without one, the others would all fall. Moreover, I also mixed them up so the Omanis and Baluchis were on the same team, rather than being separated. I told them, 'We are now Recce Family whether we are Baluchi, British, Omani or Zanzibari.' After this, I certainly had no noticeable issues between my men, while we also tried to protect Oman from the Marxist threat.

However, in my life, there has been no greater example of the importance of strong relationships than during my many polar expeditions. In subzero temperatures of below minus 50, while pulling a sledge weighing over 450lbs for hundreds of miles on end, it was important that I be accompanied by people I not only could trust, but also liked. Unsurprisingly, such arduous conditions could make one very cranky. Little incidents, such as a dropped biscuit crumb, could spark resentments that lasted for days. If I were to have embarked on an expedition with a team-mate whom I detested, then I don't think we could have lasted long. Of course, my late wife, Ginny, was also a crucial part of my expedition team, taking charge of communications, as well as everything else in between. Our love for each other, and shared determination to reach our goals, inspired us both to push ourselves beyond all reason. I felt most proud when Ginny became the first woman ever to receive the coveted Polar Medal from Her Majesty the Queen in recognition of her vital work. For this reason alone, I believe the decision to fill the Band with lovers to be inspired, for it was Ginny who certainly inspired me to keep going in my darkest hours. However, if the Thebans were to defeat the Spartans, having a devoted team of warriors was just the first step.

Pelopidas now had to train these 300 part-time warriors so that they were a match for the most fearsome military unit of their time. Stationed at Cadmea, they lived together, trained together and fought together, with everything done to ensure that they worked as a team, rather than individuals. Building strength and endurance was vital, and involved the Band running for hour after hour, in the baking-hot Greek sun, as well as engaging in wrestling matches to increase their power and agility. Their training also included intense sword and spear drills, as well as how to form a phalanx, just like the Spartans.

By the summer of 378 BC, the time had almost come for the Sacred Band to prove their worth. After just six months of training, a 20,000-strong Spartan army marched on Thebes to reclaim its territory and strike down democracy. At its head was the Spartan king, Agesilaus.

Pelopidas knew his men were not yet strong enough, or disciplined enough, to match the Spartan phalanx. Like a game of rugby, the team that had the most powerful scrum could drive the other off the field. Pelopidas needed to buy time, and this is where Epaminondas came into his own.

Everyone knew of the Spartans' legendary battle against the Persians at Thermopylae. But Epaminondas was aware that the Spartan phalanx had only worked because it had been deployed on flat, open terrain. Using this knowledge, the Sacred Band, and their Athenian allies, stood on high ground as the Spartans approached Thebes. If the Spartans wanted to fight, then they would have to do so on their terms.

On seeing the Theban formation, Agesilaus halted his men. He was also aware that his phalanx was useless when deployed on a slope. To be effective, he needed to entice the Thebans down to the flat ground. As such, he sent a few skirmishers to test the Theban and Athenian lines, only to find them easily dispatched. Still, he remained undeterred. While he couldn't deploy his phalanx, and knew the Thebans held the advantage on the high ground, he still felt confident that he could break their lines by ordering the whole of his army to descend upon them. This tactic had previously proven successful in the 394 BC Battle of Coronea, and he had no reason to doubt its effectiveness against what he believed to be a bunch of reservists. However, the Sacred Band and its allies were ready for this.

As the Spartans marched towards them, the Theban forces pointed their spears upwards instead of towards the enemy, and propped their shields against their left knee instead of hoisting them at the shoulder, ready for battle. This totally threw Agesilaus. Unsure if he was sending his men into a trap, he ordered his army to stop. A standoff ensued, with the two armies just 200 metres apart, neither making the first move.

Agesilaus was the first to blink. He subsequently ordered his forces to back down and leave the battlefield. This was clearly a situation that required more thought than he had envisaged.

Thebes not only survived another day but this crucially gave the Sacred Band a psychological advantage. While they were not ready to meet the Spartans in battle, they could at least hold them at bay with mind games. This dynamic served them well for a few years but eventually the time came when they had no choice but to fight.

When the Spartan garrison defending the nearby city of Orchomenus momentarily left its post, Pelopidas decided it was too good an opportunity to turn down. Assembling the Sacred Band, they moved on Orchomenus only to find that the garrison had unexpectedly returned. Vastly outnumbered, Pelopidas was also not convinced the Band was ready for open battle. As such, he ordered his men to return to Thebes.

However, as they did so, Spartan forces, numbering between 1,000 and 2,000 troops, suddenly emerged before them. The Thebans were outnumbered, totally unprepared, and had no way to escape. According to Plutarch, at this one of the Sacred Band said to Pelopidas, 'We are fallen into our enemy's hands.' Looking him square in the eye Pelopidas replied, 'And why not they into ours?' It was finally time to fight.

With a bloodthirsty roar, the Spartans charged towards them. Yelling above the noise, Pelopidas quickly ordered his men into a tightly packed unit, just as he had trained them to do month after month, year after year. This was something the Spartans were not expecting. They still thought they were fighting a bunch of part-time peasants, not an elite unit. Suddenly, the Sacred Band's spears had torn through their ranks and killed their leader. At this, Pelopidas showed no mercy. The Sacred Band kept attacking, moving relentlessly and viciously forward. Stunned, and in total disarray, the surviving Spartans ran for their lives.

Although the battle was small, it was still a victory to cherish. The Spartans had never before been beaten by a smaller company than their own. Now the Sacred Band was confident it could go face to face with the Spartan army, if not yet the 300 itself. But Epaminondas wasn't so sure. He still thought there was work to be done and relentlessly analysed the Spartan phalanx for weaknesses to exploit.

Through his own observations, and spies, he learnt that the Spartan army was not what it once was. By this time, there were perhaps fewer than 10 per cent actual Spartiates within its ranks. A true Spartiate was said to be someone who was born in Sparta and possessed enough land to be able to contribute to the state, and could spend their own time training to become a warrior. But, with each generation, the possession of land that entitled Spartans to citizenship, and allowed them the time to train, had been whittled away by laws of inheritance. Sparta was one of the very few places in the ancient Greek world where both men and women could inherit equally. Over time, this division of ownership meant fewer and fewer men inherited enough land to qualify and, as such, they could no longer devote their time to their training. This huge demographic shift now threatened Sparta's long-standing dominance on the battlefield.

To fill the void, their vast army became increasingly dependent on non-Spartan allies, known as Perioeci. But these troops were not always reliable. Essentially made up of slaves, or farmers obliged to fight for their Spartan overlords, they had no real loyalty to the Spartan cause. This was a fault line that Epaminondas thought he could exploit. If the Thebans could somehow break the Perioeci's allegiance, then the whole Spartan army would collapse. To do this, Epaminondas realised he had to show the Perioeci that the feared Spartan 300 could be defeated.

As detailed by the Athenian historian Thucydides, and as seen at Thermopylae, when the Spartan phalanx adopted a defensive position, they overlapped their shields, which created an impregnable wall that they hid behind. To protect themselves, a few soldiers in the line tended to move their shields slightly towards their right. This saw the soldiers to their left have to do the same in order to protect themselves, which caused a chain reaction down the line, with those towards the far right now the most unprotected. It was for this reason the Spartan generals put their best soldiers on the right of the phalanx. But Epaminondas thought he could exploit this. By placing the Sacred Band on the left, to face the Spartan right, a shock move could quickly overcome them and cause the Spartan phalanx to collapse. Upon seeing Sparta's best warriors and leaders quickly perish, Epaminondas believed this would so terrify the Perioeci, who didn't want to fight for their overlords anyway, that they would all flee or surrender.

However, even Epaminondas was aware that, despite all of the Band's training, and their bond, it might not be enough to overcome the very best warriors Sparta had to offer. He therefore needed to find one more edge before he could put his strategy to the test.

Once more, he analysed the Spartan formation and realised that it was never stacked more than twelve ranks deep. If he could stack fifty ranks behind his Sacred Band on the left, then they might just be able to overwhelm the formidable Spartan right, swarm the king and his commanders, and hand Sparta its first defeat in open battle in over a century. The Macedonian author Polyaenus, in his *Stratagems of War*, stated that, in effect, Epaminondas wanted to catch a snake, crush its head and show his men how useless the rest of it was. The Battle of Leuctra in 371 BC would give Epaminondas a chance to test his theory.

With age and illness apparently getting the better of King Agesilaus, Sparta's other king, Cleombrotus, ordered his vast army of 11,000 Spartans and Perioeci to take Thebes once and for all. Marching to the plain of Leuctra, they were met by Pelopidas and his Sacred Band, as well as 7,000 troops. On this flat ground, Cleombrotus expected his phalanx to record an easy victory and to put an end to the Theban rebellion for good. It was clear to Epaminondas that his men were vastly outnumbered. If his plan did not get results quickly, then they would all be slaughtered. The Sacred Band's attack needed to be one of brutal shock and awe if it was going to work.

Just as Epaminondas expected, the Spartan king, his commanders and his best warriors lined up at the front, on the right of the phalanx, eight ranks deep. In response, he quickly ordered a bulk of his formation to assemble on the left, behind the Sacred Band. Cleombrotus could see what was happening but before he could act Pelopidas gave the signal to attack. At this, the Sacred Band launched themselves into the right of the Spartan phalanx, with over fifty ranks of shields and spears behind them.

Twenty of the Sacred Band were instantly killed by Spartan spears, but nonetheless they stuck tight together, the weight of their ranks ploughing them forward, causing the Spartan wall to crack under the strain. As gaps suddenly opened, carnage erupted. Spartans were stabbed and slashed, while Pelopidas tore through the bodies and grabbed King Cleombrotus. Stabbing him again and again, Pelopidas left nothing to chance, ensuring that Cleombrotus was the first Spartan king to be killed in battle since Leonidas. The king's death sent a shockwave through the rest of the Spartan army. Not only had the 300 been defeated but their king was dead, along with over 400 of the 700 Spartiates.

As Epaminondas predicted, the slaves and allies of Sparta now decided to retreat. His strategy was a triumph, representing the most shattering defeat in the history of Sparta, and signalling the end of Spartan dominance on the battlefield. Since the number of Spartiates was already small, the death of over half of them saw the Spartan army gradually disintegrate. In its absence, the Sacred Band became the most elite military force in Greece, while they had also ensured that Thebes remained free.

However, battles with the weakened Spartans would continue for a few years yet, with Epaminondas fatally wounded at the Battle of Mantinea in 362 BC. As his ailing body was carried back to the safety of the Theban camp, he asked his colleagues if the Thebans had been victorious. Upon being told that his tactics, similar to those at Leuctra, had again brought victory he smiled in response and uttered, 'Then it is time to die.' When one of his friends said, 'You die childless, Epaminondas,' he replied, 'No, by Zeus, on the contrary. I leave behind two daughters, Leuctra and Mantinea, my victories.'

In Epaminondas's absence, the Sacred Band would soon come up against another formidable foe. As the might of the Macedonian army moved on Thebes, a prodigious 18-year-old led the cavalry, intent on marking his name in history ...

4

ALEXANDER THE GREAT
AND THE SOGDIAN ROCK

338 BC

On a bright August morning, two armies faced off against each
other in the Greek town of Chaeronea. Once again, the fate of
Greece hung in the balance. This time, it was the rising power of
Macedonia that sought to rule, and all that stood in its way was
the combined forces of Thebes and Athens.

The king of Macedonia, Philip II, was well aware of the military strength of Thebes. In 367 BC, the Thebans had taken him
prisoner, where he not only received a military education from
Epaminondas, but it is also believed that he became a lover of
Pelopidas. Both relationships gave Philip a rare opportunity
to study how Thebes's previously amateur army had become
Greece's most powerful armed force. In 359 BC, after returning
to Macedonia, Philip's ascent to the throne was endangered by
wild barbarian tribes to the north and wily Greek cities to the
south. Taking inspiration from the Sacred Band, he rearmed and
retrained his infantry, fending off enemies in his own country, and
then looked to conquests further afield. More often than not, it
was the Macedonian phalanx that proved the difference. Inspired

by the Sacred Band, Philip had also made a few improvements of his own.

Firstly, he had replaced the outdated hoplite spear with the sarissa, an 18–20ft pike, to give his men further reach. However, this was just the beginning of his phalanx revolution. As his phalanx moved into formation, every man would carry their sarissa upright, and then, just before they were to engage, the first five ranks would lower their deadly pikes horizontally, creating a deadly wall of iron. This would set off a chain reaction as the lines just behind would then lower their own pikes at a 45-degree angle, while the rows further back would keep their pikes upright until they were engaged. Thanks to the great length of the pike, four deadly sarissa heads could protrude ahead of the first infantryman in the phalanx. This gave Philip and his infantry a great advantage. The sheer offensive power of multiple advancing sarissa could steamroller any opponent. Yet Philip also ensured his phalanx was flexible. Formed in line or in column, with varying depth to suit the circumstances, it could be packed close or strung out thin, or in a wedge formation to penetrate an enemy's front. Either way, he was a master at utilising it in any situation, ensuring his phalanx was considered the deadliest in the ancient world.

With his well-trained and well-armed phalanx, the Macedonians crushed the elite infantry of both the Illyrians and the Greeks, as well as the hardened warriors of Thrace and Paeonia. However, while his phalanx and infantry were formidable, Philip had another secret weapon at his disposal: his son Alexander, the 18-year-old commander of his elite cavalry.

Several legends surround Alexander's birth in 356 BC. According to Plutarch, his mother, Olympias, had a dream on the eve of the consummation of her marriage to Philip, in which her womb was struck by a thunderbolt. Sometime after the wedding, Philip also had a dream in which he secured his wife's womb with a seal engraved with a lion's image. Thereafter many Greeks believed that Alexander's virgin birth proved him to be the son of their god Zeus. Alexander certainly didn't believe his pedigree was a mere genealogical fiction. As he grew older, he truly believed he hailed from the gods, and, in later years, he behaved explicitly as though he were the lineal descendant of both Hercules and Achilles.

Philip also believed his son was special. On the day of Alexander's birth, Philip received news that his forces had defeated the combined Illyrian and Paeonian armies, while his horses had also won races at the Olympic Games. These good omens were, however, just a taste of the glory Philip's young son would soon bring him, and his country.

Educated by Aristotle, at the Temple of the Nymphs at Mieza, Alexander developed relationships with children of Macedonian nobles, such as Ptolemy, Hephaistion and Cassander. In time, these young boys would become Alexander's generals in his elite special forces, known as the 'Companions'.

Hailing from the upper classes, they were able to acquire and maintain the best armour and horses, and were experts at wielding the spear and shield. Eventually they would become the first shock cavalry in history. In battle, Alexander would wait for the Macedonian phalanx to pin the enemy in place, before unleashing his Companion Cavalry on the enemy's flank, or behind, with lightning speed and ferocity.

In 338 BC, Alexander's Companion Cavalry would face its greatest test yet. Outside the town of Chaeronea, they faced Greek forces, including that of the Sacred Band. Sadly, details of the battle itself are scarce, with Diodorus providing the only formal account. However, he recounts that 'the battle was fierce and bloody' and that 'victory was uncertain' until Alexander entered the fray. Riding at the head of his Companions, and with the battle in the balance, he had 'his heart set on showing his father his prowess'.

From here, Alexander led the charge of his Companions, and 'was the first to break through the main body of the enemy, directly opposing him, slaying many; and bore down all before him – and his men, pressing on closely, cut to pieces the lines of the enemy; and after the ground had been piled with the dead, put the wing resisting him in flight'.

The 18-year-old boy wonder had just delivered the most crushing defeat, on the biggest stage, of his young life. Not only had he proven his prowess as a warrior, but also as a leader. In front of him now were the spoils of his victory – thousands of dead Athenians, as well as the bloodied bodies of the Sacred Band.

Plutarch, in his *Life of Alexander*, wrote of the young commander's courage, '... he is said to have been the first man that charged the Thebans' sacred band ... This bravery made Philip so fond of him, that nothing pleased him more than to hear his subjects call himself their general and Alexander their king.'

After Alexander's famous victory, Athens was forced into an alliance with Macedonia, while Thebes lost its rich agricultural lands in Boeotia, with the Sacred Band no more. According to Plutarch, Philip was extremely moved by the Band's courage, having of course once been so close to their founders, Epaminondas and Pelopidas:

> After the battle, Philip was surveying the dead, and stopped at the place where the 300 were lying, all where they had faced the long spears of his phalanx, with their armour, and mingled one with another, he was amazed, and on learning that this was the band of lovers and beloved, burst into tears and said, 'Perish miserably they who think that these men did or suffered aught disgraceful.'

So touched was Philip at the sacrifice of the Sacred Band that it appears he built a monument to them, which was discovered by a

British tourist in 1818. While out horseback riding near the battle-field, the tourist tripped over a rock that, upon some digging, was revealed to be a massive stone lion. Inside the monument, 254 skeletons were later discovered, believed to be those of the Sacred Band, with the lovers wrapped in eternity to each other.

The Macedonians' victory at Chaeronea was a turning point in history. With the Greeks conquered, Philip now turned his military ambitions to Persia. However, before this could be realised, he was assassinated, with Alexander coming to the throne at the age of just twenty.

Despite his youth, Alexander launched one of the most daring military campaigns ever attempted. With just 50,000 men, he set out to achieve his father's dream of conquering the vast Persian Empire. By now, it stretched from modern-day Turkey in the west and included most of the Middle East, an area of 3 million square miles.

Nevertheless, Alexander commenced his invasion in 334 BC and, over the course of four years, won a series of decisive battles, most notably the Battle of the Granicus River and the Battle of Issus, which finally marked the end of the Persian Empire. After his men had killed or captured over 100,000 Persians in Issus, the Persian King Darius offered Alexander half his empire and marriage to his daughter, in an attempt to reach a truce. Alexander replied that he already had half of the empire, and intended to take the rest.

Such conquests saw Alexander heralded as a military genius, whose guile, ingenuity and lateral thinking allowed him to defeat far superior forces. Many started to believe that the prophecy might actually be true: Alexander really was descended from the gods. His men vowed to follow him to the end of the earth, which is exactly where he wanted to take them, as he endeavoured to reach the 'ends of the world and the Great Outer Sea'.

In 1979, after seven years of planning, my colleague Charles Burton and I embarked on the 'Transglobe Expedition'. The aim

was to become the first expedition to make a circumpolar navigation, travelling the world 'vertically', traversing both of the poles
using only surface transport. In effect, we were looking to reach
both 'ends of the world'. It took us over three years and 100,000
miles, travelling across the Sahara, through the swamps and jungles
of Mali and the Ivory Coast, over huge unexplored crevasse fields in
Antarctica, through the inhospitable Northwest Passage, and into
the unpredictable hazards of the Arctic Ocean. In temperatures
that veered between 40 degrees in the Sahara and minus 50 in the
Antarctic, we suffered countless injuries, from frost-nipped noses,
fingers and toes, to blisters and sores down to the bone, not to
mention facing the threat of polar bears. Alexander might very well
have achieved all of this and more, but I'm not entirely sure the poles
would have reached his expectations for the 'ends of the world'.
All I can say is it was very white and very cold. However, the mere
thrill of getting there, which was certainly one of the reasons that
spurred me on, might just have been enough for Alexander. After all,
sometimes it is the journey, rather than the destination, that counts.

In any event, before he embarked on his ambition to reach the
end of the world, Alexander soon had his next conquest in sight:
India. This exotic country was one that not even his hero Hercules
made his own. Moreover, it was all that was left for Alexander to
conquer in the ancient world.

However, standing in Alexander's way, between Persia and India,
were the tribal badlands of Bactria and Sogdiana, what we know
today as Afghanistan and Tajikistan. This was one of the world's
most difficult places to control, as the Soviet Union and United
States have found to their cost in recent times. Warlords in the
regions knew the land better than anybody. Disappearing into the
vast mountains, they then picked their enemy off from the shadows.

By now, Alexander's supply chain ran thousands of miles back
to Greece. He depended on this to move troops, horses and gold,
all vital when conquering a country. This was something the

Afghans understood, and they hit the chain at every opportunity. And it was never-ending. Even if Alexander killed one warlord it seemed another dozen took his place. After three years of brutal suppression, he had done all he could to overcome this threat, but one warlord still stood in his way: Oxyartes.

The Greek historian Arrian of Nicomedia detailed in his book *The Anabasis of Alexander* that Oxyartes retreated with his tribe and his precious 16-year-old daughter, Roxana, to the Sogdian Rock, a mountain fortress located north of Bactria. Twelve miles in circumference, it rose to a summit some 3,000ft up. From here, Oxyartes could still swoop down on Alexander's supply lines while remaining untouchable. Alexander knew he couldn't take India until Oxyartes had been dealt with, but with his impregnable mountain fortress it looked an impossible task. The intelligence Alexander received was also unfavourable.

While Alexander's army outnumbered Oxyartes' ten to one, he was told that any attempt to face them head-on would be a suicide mission. The only route to the fortress was a narrow winding footpath, which was out in the open. Should his troops attempt to storm the castle, they would either be picked off by archers or forced downhill. It appeared the only option was to lay siege, but Alexander was also told that Oxyartes had enough supplies to last for months, while there was a never-ending supply of fresh water, thanks to the snow on the mountain.

Alexander had no patience for a long siege. In order to break the impasse, he sent an envoy to Oxyartes promising to give the warlord, and his tribe, safe passage if they surrendered immediately. Such was his confidence in this impregnable fortress that Oxyartes mockingly told the envoy that Alexander would need 'men with wings' to capture him. This response infuriated the Macedonian king. But, as Oxyartes' taunt played over in his mind, it soon gave him an idea.

Quickly, he ordered his commanders to muster 300 of the

best climbers in his army. Many of these men were either former mountain herdsmen, used to scaling rugged cliffs, or members of his crack infantry unit, known as the shield bearers, who were always in the first wave of attack over high enemy walls. During their duty, all had been involved in dangerous situations, but even they were astonished when Alexander told them of his plan.

Scaling the 3,000ft rock, at night, without making any noise to alert the enemy, even if they should fall to their deaths, Alexander wanted them to reach a crag overlooking the rebel base. He was convinced that their sudden appearance would deliver such a shock that the rebels would surrender without a fight. After all, Oxyartes believed that only men with wings could reach the summit, and that's just what Alexander wanted him to believe.

Unsurprisingly, many of the men were apprehensive. Climbing the rock at night seemed suicidal, especially in such cold and snowy conditions. What's more, they wouldn't be able to climb 3,000ft in one single night. This meant that, when daylight hit, they would need to hide on the mountain, all day, until darkness descended and they could then resume their climb. In order to climb such a mountain, they also needed to be as light as possible. This would mean taking with them only minimal clothing and provisions, surely not enough to sustain at least two days of arduous climbing. Although one essential item Alexander insisted they all took with them was a white sheet, which they were to wave as a signal once they had reached the top. This reminds me of my time at Eton, where for some reason I became addicted to stegophily, the official name for the sport of climbing up the outside of buildings, particularly at night, and leaving items on their topmost spire, dome or lightning conductor. However, one such escapade almost saw me expelled from school after my climbing partner and I had affixed a flag to the 'summit' of School Hall's dome. I was lucky enough to escape in the ensuing chase but, alas, my poor partner was caught and asked to leave at the end of term. Such punishment nonetheless paled in comparison to what would happen to Alexander's mountain men if they were to be caught.

Alexander was, however, prepared for his men's scepticism. He therefore offered them a sweetener. The first man to reach the summit would be rewarded with a huge prize of twelve talents (the equivalent of $250,000 today), while the second man would receive eleven talents, the third ten, and so on to the twelfth, who would receive 300 gold darics.

It might have seemed like a suicide mission but Alexander's mountain men couldn't turn this down. The dangers were clear. They could die from exposure, fall to their deaths, or be picked off as sitting targets by enemy archers, but it could also make them very rich men. Moreover, it could also put an unknown man into the good books of the world's greatest conqueror, a valuable position indeed.

Like many of you, I am not sure I would have been tempted by such a proposition. I have always had a tremendous fear of heights and this mission would have terrified me. Over the years, I have tried to confront this fear head-on, with little success. My first attempt saw me suffer heart issues while climbing Everest, which meant I had to turn back. While I would go on to conquer the world's highest mountain a few years later, I found that it still hadn't really solved anything. Climbing Everest was, of course, a monumental task but it didn't really involve scaling sheer cliff faces, with a massive drop below. As such, I turned to the supposed daddy of vertigo-inducing climbs – the north face of the Eiger mountain in the Alps.

The 6,000ft of vertical limestone and black ice had killed over forty of the world's top climbers. Joe Simpson, of *Touching the Void* fame, has written of this so-called 'Murder Wall': 'It wasn't the hardest or the highest. It was simply "The Eiger". The very mention of the name made my heart beat faster. The seminal mountain, a metaphorical mountain that represented everything that defines mountaineering – a route I had dreamed of climbing my entire adult life.'

My task was made slightly harder owing to the fact that some of the fingers on my left hand had previously been amputated due to suffering frostbite on an expedition. Thankfully, they were not to cause me too many problems. One thing that did, however, terrify me was when one of my boots skidded off the rock, and I was left hanging off the mountain by just the axe in my right hand and the smallest of footholds. My mouth felt very dry after that, and I still faced two more days, with yet more slips and falls prevented only by my axe precariously buried in the rock. At times, I had to avoid looking down and occupy my mind with all manner of things, so that no tiny chink of terror could burrow into my psyche and render me a gibbering fool. When I accidentally dropped my windproof jacket, and saw it fall below, the subsequent butterflies in my stomach threatened to eat me alive.

Night-time was also a testing experience. After climbing all day, I was exhausted. However, rest would not come easily. Each night we had to find a ledge upon which to camp, which was often no more than 4ft wide. After clipping my harness to a rock bolt, I would lie with my nose right up against the wall, and my backside protruding over the edge of the ledge, with the void thousands of feet below. I daren't close my eyes for fear that one wrong move would see me dangling over the edge. That was certainly a jolt that would have woken me up.

While I managed to 'conquer' the Eiger, I still had not conquered my vertigo. It seems that it is a condition I will be stuck with, but I certainly can't be accused of hiding from it. In any event, the idea of climbing the Sogdian Rock, like Alexander's men, without any equipment, is certainly one that would make anybody think twice.

As darkness enveloped the Sogdian mountain, the men began their preparations. With at least two nights of climbing ahead, they had to travel light. In order to cut down on the quantity of rations they took with them, they ate and drank as much as they could before they set off. They also wore minimal clothing, despite it being a cold spring night. With a sheer rock climb, any added weight could be the difference between success and failure. And in this case failure meant death.

In the darkness, the men soon started to scale the mountain, desperately looking for secure hand- and footholds, with every move a life-or-death decision. Knowing they couldn't make any noise, the men were unable to hammer any metal spikes into the rock face to secure a rope or help provide added grip. In their place, they carried rocks of various sizes, which they lodged into cracks. These could help to either secure a rope or act as a foot- or handhold. Some, however, preferred to free climb. This not only cut down on noise, but also time. Yet it was incredibly dangerous at the best of times, let alone in the pitch-black. It is a prospect that brings me out in a cold sweat, although at least in the darkness you would not be able to see the drop below.

But Alexander's mountain climbers soon faced a new threat. Already exposed to the elements, they were before long lashed by freezing rain. With their hands quickly becoming frozen, the rocks also became perilously wet, making it almost impossible to find a grip. Feeling blindly around for something to hold, many lost their grasps and fell to their deaths. As promised, those brave souls fell without so much as a scream to avoid alerting the enemy. Meanwhile, their surviving comrades continued upwards, inch by inch.

From his base at the foot of the mountain, Alexander received the casualty reports. By dawn, a dozen men had fallen to their deaths. This saddened him but he at least still had 288 men climbing the rock face. But another test now arrived. With daylight breaking, the men had to hide on the rock all day, waiting until the sun set before they could resume their perilous climb. Without the cover of darkness, they were dangerously exposed to the elements, as well as to being detected by Oxyartes' sentries. For Alexander's plan to work, the element of surprise was crucial. Hiding in nooks and crannies, they used the snow to hydrate themselves, while waiting for nightfall.

Thankfully, the day passed without Oxyartes seeing, or even suspecting, that Alexander's men were climbing the mountain. As darkness descended, these tired, hungry and thirsty men again began to climb. However, while they might be getting closer to the summit, they were also getting closer to the enemy, who significantly outnumbered them. If they should be spotted, they would be hit by a shower of arrows and spears. And if they were to be captured, torture and death would await. But still they continued on, motivated by Alexander's promised prize.

Upon rising the following morning, Alexander looked to the summit for any white sheets. There were none. All his men looked to have perished in the night. This represented a tremendous setback and put all his invasion plans in jeopardy. However, from the base of the Sogdian Rock, Alexander suddenly saw a white sheet fluttering in the wind, then another, and another, as more of his warriors reached the towering summit. Incredibly, 270 of his 300 men had defied the impossible odds and made it to the top.

There was now no time to waste. With this signal, Alexander launched his double-edged attack. His ground forces swiftly advanced up the rock, just as Oxyartes and his rebels noticed the threat from above. The warlord was astonished to see Alexander's men surrounding the base from above and below. He truly believed that Alexander's soldiers must have wings. At this sight, the confidence drained out of him and he surrendered without a fight, just as Alexander had hoped.

Finally, Alexander had control over this rebellious region and could make his move into India. However, he realised that to maintain this control he now needed to change tack. Conveniently, Oxyartes' daughter, Roxana, was described as one of the most beautiful women in the land, and Alexander fell in love with her at first sight. He also realised that, if they were to marry, this would give him a strong family foothold in the region. Thankfully, Roxana agreed to his request for marriage. To further cement this foothold, Alexander appointed Oxyartes as the governor of the Hindu Kush region, which adjoined India. His one-time enemy now had a stake in his failure and success. It was another brilliant move, as Oxyartes persuaded other warlords to rally behind his new son-in-law.

Alexander's victory at the Sogdian Rock cleared the way for his invasion of India in 327 BC, where he won an important victory at the Battle of the Hydaspes. However, as his forces marched east, across the Punjab, and reached the Hyphasis River, his troops finally mutinied, worn out from years of conquests, refusing to go any further. Alexander reluctantly accepted the end of his monumental quest and began the extremely dangerous journey back to Macedonia.

On his journey home, he was badly wounded by an arrow, which lodged in his chest. Despite surviving this injury, he would finally perish in June 323 BC, at the age of thirty-two. Historians are unclear as to the cause of his death. Some believe it was a disease or infection such as malaria or cholera. Others suggest he was poisoned by an assassin, while some believe he might have even committed suicide after becoming depressed.

Nonetheless, within less than two decades, Alexander had truly earned the title of the most dangerous man in the world. In his capacity as king, commander, politician, scholar and explorer, he led his army over 11,000 miles, founding over seventy cities and creating an empire that stretched across three continents and covered around 2 million square miles. That he never lost a battle, despite typically being outnumbered, is not only testament to his leadership but also to the strength of his elite special forces, the Companion Cavalry, and the mountain men of the Sogdian Rock.

However, Alexander's death soon sparked a bitter power struggle. At the time, Roxana was pregnant with Alexander's only heir. As such, she decided to ensure her unborn son had a clear path to the throne, murdering Alexander's other wife, Stateira, as well as her sister and her cousin. Upon her son's birth, Roxana named him after his father and it seemed that he would follow in his father and grandfather's footsteps. Yet the power struggle continued to rage, culminating in Roxana, and the young Alexander, being murdered by four Macedonian generals. The generals subsequently carved up separate parts of the empire between them and ruled until the coming of the Roman legions some 300 years later. And while Rome would have its own special forces to thank for furthering its empire, they were also intent on destroying it from within, as not every elite special force always acted in the best interests of their leader . . .

5

THE ROMAN
PRAETORIAN GUARD

AD 41

As the Roman emperor Caligula sat on his throne in the imperial palace, watching a series of games and dramatics, his elite imperial guard surrounded him on all sides. Sporting their famous Attic helmets, body armour and rectangular scutum shields, they sent out the message to the large and boisterous crowd that to get to the emperor they would first have to go through them. From the death of Tiberius Gracchus in 133 BC through to the assassination of Julius Caesar in 44 BC, an event that effectively ended the Roman Republic and saw Augustus become Rome's first emperor in 27 BC, Rome was always full of conspiracies, and there was no shortage of individuals to carry them out. It was for this very reason that Augustus had been inspired to create the Praetorian Guard.

By the time Augustus had come to power, the Roman world stretched to the Middle East and included much of western, central and southern Europe, as well as north Africa and Turkey. Unsurprisingly, the head of the Roman Empire was always a target and required protection at all times. Indeed, Augustus knew this more than most. While his life was always under threat in times of conflict, it was even in danger when walking the streets of Rome. In 39 BC, he had been stoned by starving rioters when their grain supply had run low and had only just escaped with his life thanks to the help of a few attendants.

Forming a personal bodyguard was hardly revolutionary for a man in Augustus's position. Over the preceding centuries, those in high positions in Rome had always employed some form of guard. However, the Praetorians were a far more permanent, regulated and co-ordinated force than anything that had gone before. Serving with one purpose, to protect the emperor and his family at all times, they would not only be at his side but they also sought out, and crushed, any threats of rebellion. With his rise to power heralding the end of the republic, Augustus needed to assert his authority and have a guard he could trust to carry out his demands. While his rule was never explicitly said to be a military dictatorship, the Guard allowed him to maintain power with an iron grip, while also putting on the charade that he was overseeing a democracy.

However, as the Guard grew, with Cassius Dio estimating there to be as many as 10,000 by AD 5, their duties became ever more haphazard. Staffing prisons, and collecting taxes, it was said that sometimes they also acted as Rome's fire brigade, with the emperor dispatching his personal guard to blazes in the city so that he could show the citizenry he was concerned for their welfare. On the tomb of one Praetorian Guard, Vinnius Valens, it was even recorded that he would hold up wine carts while they were emptied or use his strength to force wagons to come to a halt with one hand. The Guard clearly occupied a number of positions in Rome, and came to take on utmost importance. However, its primary duty was always to ensure the emperor's safety.

On this mild winter's day, Caligula's life was in the hands of Cassius Chaerea, who had endured the rigorous Praetorian recruitment and training process to be trusted with such a prestigious duty. Meeting the initial requirements, Chaerea hailed from central Italy, was between the required age of fifteen and thirty-two, was in good physical condition, had a good moral character, and came from a respectable family. In addition, he also managed to obtain letters of recommendation from important leading figures in society. With this, he was based at the fortress-like Castra Praetoria, just outside the perimeters of Rome, where he lived and trained with up to 15,000 other Guards. Unfortunately, specific details regarding the Praetorians' training have been lost to time, as have many details of Chaerea's service, but we know via the ancient historian Tacitus that he had distinguished himself on the Germanic frontier with bravery and skill.

Like many before him, Chaerea had no doubt been lured to the Guards by the many benefits the role had to offer. Not only were the Guards the only individuals who were permitted to bear arms in the centre of sacred Rome, their mandatory service was far less than that required for the legions, having to serve sixteen years compared to twenty. Moreover, their pay was substantially higher, said to be at least double the pay of a legionary. This was further augmented by the *donativum*, a sum awarded to the Guard by each new emperor. This could amount to the equivalent of several years' pay, and was sometimes repeated at events of great importance. Entering the Guard also presented the likes of Chaerea with a route to the top. A Guard could climb the ranks, going as high as the Praetorian Prefecture, whereby he would serve as a representative of the emperor. Some Guards even entered politics upon their retirement, with their eye on succeeding the emperor himself. Unsurprisingly, the Praetorians regarded themselves as a cut above the rest of the Roman military machine, and it gave the men involved considerable standing in their communities.

Yet Chaerea's job was not an easy one. Guarding the emperor was always a challenge but even more so when it came to the likes of the volatile Caligula. Rude and demanding, the emperor never missed an opportunity to mock Chaerea's voice and even gave him passwords that were designed to shame and embarrass him. Suetonius has reported that whenever Caligula had Chaerea kiss his ring, Caligula would 'hold out his hand to kiss, forming and moving it in an obscene fashion'. And yet, despite taking every opportunity to insult the man guarding his life, Caligula actually had the Praetorian Guard to thank for his ascension to the throne.

Caligula's father, Germanicus, was the nephew, and adopted son, of Augustus's successor, Tiberius. Nevertheless, he was designated as Tiberius's heir, even above his own son, Drusus. Tiberius recognised that Germanicus not only came from the bloodline of Augustus, but was also an acclaimed military leader, playing a crucial role in repressing a mutiny by the armed forces in Germania. His loyalty to Tiberius was also without question.

However, Sejanus, Tiberius's Praetorian prefect, also had his eyes on the throne. To get it, he would need to make a series of power plays to get rid of his rivals. Firstly, he falsely informed Tiberius that Germanicus, father of Caligula, was plotting against him. Enraged by this betrayal, Tiberius poisoned the man he had once earmarked as his heir. Caligula's mother sought revenge against Tiberius but she was detained and imprisoned, along with her sons. However, due to Caligula's young age, Tiberius opted to spare him and instead sent him to live with his mother.

With Germanicus and his family out of the way, Sejanus proceeded to kill Tiberius's next heir – his son, Drusus. Following Tiberius's withdrawal from Rome in AD 26 to reside in Capri, Sejanus was left in charge, with Tiberius calling him *socium laborum*, 'my partner in labours'. It appeared Sejanus's plan had worked. He was now next in line to the throne.

At this, Sejanus established the Praetorian Guard within Rome itself, at the newly built Castra Praetoria. While this allowed him ever more control over Rome's most elite military force, he also wanted all of Rome to know it. To prove this, he ordered the Praetorians to put on an exhibition of their drill in front of watching senators. This display put them all on warning that Sejanus was in charge. Meanwhile, he controlled access to Tiberius, ensuring that he only passed on the messages he liked, and ignored those he didn't. Sejanus's power in Rome was soon unchallenged by any heir or rival and it was only a matter of time before he secured the greatest prize of all.

However, when Tiberius discovered Sejanus's plotting in AD 31 he ordered him to be executed. His body was subsequently thrown down the Gemonian Stairs, which led down from the Capitoline Hill to the Roman Forum, while his children, and any known associates, were also killed.

By this time, the only surviving heir to the throne was Caligula, with Sejanus having eliminated all other rivals. Caligula was subsequently sent to live with Tiberius on the island of Capri, where he was adopted by his father's killer. Tiberius, however, had no illusions about his heir, stating, 'I am bringing up a viper for the Roman people and Phaethon for the world.' When Tiberius passed away in AD 37, Caligula became emperor of Rome at just twenty-five years of age, chillingly telling his grandmother, 'Remember, I can do anything to anybody.'

Despite this, Caligula initially proved to be immensely popular. He granted bonuses to those in the military, including the Praetorian Guard, eliminated unfair taxes and freed those who were unjustly imprisoned. He also ensured Rome was never short of entertainment, laying on lavish chariot races, gladiator shows and plays, while also building a new racetrack known as the Circus of Gaius. Infrastructure also received heavy investment. New roads were built, existing roads repaired, and Caligula's improvements to the harbour at Rhegium and Sicily allowed increased grain imports from Egypt, so that his people would never go hungry. The aqueducts, Aqua Claudia and Anio Novus, were also considered engineering marvels of their time.

However, at the imperial palace in AD 41, those achievements now seemed a distant memory. While he was surrounded by Chaerea and his fellow Praetorians, Caligula also had a team of guards mingling with the crowd, wearing simple togas in order to blend in. Known as *Speculatores*, they acted as Caligula's secret police, listening out for any criticism of the emperor and, if necessary, arresting those they believed to be a threat. And on this day, they had their work cut out. During the proceedings, some in the crowd took the opportunity to beg Caligula to lighten their burden a little by remitting taxes. Rather than address these issues, Caligula instead nodded to his *Speculatores*. Moments later they had arrested any of those who continued to shout at the emperor and dragged them kicking and screaming towards their deaths.

At the sight of this, most chose to keep their opinions to themselves. They were well aware that, at a previous games, Caligula was said to have ordered his guards to throw an entire section of the audience into the arena, where they were eaten by wild beasts. On occasion, he had also forced his Praetorian Guard to do battle with the beasts just for his own amusement. Such stories merely emphasise how Caligula's reign had descended into anarchy.

His problems appear to have begun just a few short months after taking power, when he fell seriously ill, with some suggesting he had been poisoned. While he recovered physically, his mental well-being was said to have been dramatically affected.

Following this bout of illness, Caligula suddenly delighted in humiliating those around him. A favourite pastime was to sleep with other men's wives and then brag about it, killing their husbands should they complain. Suetonius and Cassius Dio also accused Caligula of engaging in incest with his sisters and claim he prostituted them to other men.

Gripped by paranoia, Caligula also began to execute all those close to him. As his support base dwindled, he made extravagant payments to keep supporters on side but this caused Rome to suffer a severe financial crisis in AD 39. The ancient historian Suetonius, in his *The Lives of the Twelve Caesars*, claimed that, to dig himself out of this hole, Caligula began to falsely accuse, fine and even kill individuals for the purpose of seizing their estates. He was also said to levy taxes on lawsuits, weddings and prostitution. In the first year of Caligula's reign, he squandered over 2.7 billion sesterces.

Despite these financial difficulties, and with some of his people reduced to starvation, Caligula continued to spend money freely on a catalogue of outrageous whims. For instance, in response to Thrasyllus of Mendes' claim that Caligula had 'no more chance of becoming emperor than of riding a horse across the Bay of Baiae', he built a temporary floating bridge, using ships as pontoons, which stretched for over 2 miles, from the resort of Baiae to the neighbouring port of Puteoli. Upon the bridge's completion, Caligula, who could not swim, proceeded to ride his favourite horse, Incitatus, across the water wearing the breastplate of Alexander the Great. This was all at great expense, and the bridge was of course only temporary, so served no larger purpose than pandering to Caligula's ego.

Relations between Caligula and the Roman senate suddenly became strained. While the emperor executed a number of senators for treason, he made others wait on him, or run beside his chariot until they collapsed, in order to make an example of them.

From this point, Caligula's mental health took a nosedive. He not only made his horse, Incitatus, a priest but he also acquired delusions of grandeur, appearing in public dressed as various gods and demigods, such as Hercules, Mercury, Venus and Apollo. In AD 40, Caligula even announced to the senate that he planned to leave Rome permanently and move to Alexandria in Egypt, where he hoped to be worshipped as a living god. It was now clear to many senators that a new emperor had to be installed on the throne. Caligula was now a marked man, with many queuing up to do the deed.

The Praetorian Guard were soon inundated with potential plots against the emperor. While Anicus Ceralis and Sextus Papinius were executed for planning to assassinate Caligula, Aemilius Regulus was also known to be making a move. Another who wanted Caligula dead was Annius Minucianus, whose close friend, Lepidus, a man with few equals in Rome, had been killed by Caligula. It seemed that every day that Caligula ruled, the list of likely assassins was getting longer and longer.

But, as Caligula sat on his throne, such thoughts were far from his mind. Believing himself to be a god he failed to see how his subjects could not cherish and adore him, even in the face of all evidence to the contrary. Yet the games and dramatics were too dreary to amuse a budding god. Bored and restless, Caligula rose from his seat, grunted at Chaerea, and made to leave the palace, with the eyes of the nervous crowd upon him.

With Caligula heading to an underground corridor, Chaerea and his fellow Praetorian Guards followed procedure and flanked him on either side. They all knew the drill by now and had learnt to ignore their leader's unpredictable behaviour, which saw him curse at the time he had just wasted, while also lashing out at anyone who dared cross his path. As the exit loomed up ahead, some of the men glanced at their leader, Cassius Chaerea. He nodded back, placing his hand on his sword as he did so.

Suddenly, Chaerea quickened his pace. As he reached Caligula's side, the emperor turned and unleashed a torrent of insults but, before he could finish, the very man who was meant to protect his life had stabbed him in the back. Sickened by Caligula's appalling behaviour, Chaerea had agreed to kill him, with the support of the senate. With Caligula's eyes widening in shock, the rest of his guards descended on him like locusts, stabbing again and again, leaving no room for error. They were all well aware that, if the emperor should survive this assassination attempt, then their executions would not be quick or painless. But it soon turned out that killing the emperor was the easy part.

Upon hearing of Caligula's death, the military was dismayed. Not only had they been well paid under his rule, but they were also aware that the senate had arranged his death as an opportunity to restore the republic and regain power. For the military, but especially for some elements of the Praetorian Guard, this was inconceivable. If the republic should return, then the Guard itself might cease to exist. As such, many key military figures refused to support Chaerea's coup, which left him with just one option: to hunt down and kill any heirs to the throne.

Over the coming days, Chaerea and his fellow guards embarked on a murderous spree. If they could survive, while eliminating all successors to Caligula, then there would be no option but to return to a republic. Perhaps of more importance was that, with none of Caligula's heirs surviving, they also would not be held responsible for his death. They tracked down Caligula's wife, Caesonia, and murdered her, as well as her young daughter, Julia Drusilla, by smashing her head against a wall. Anyone with a claim to the throne, no matter how slight, was to receive similar treatment.

However, as a rival faction of the Praetorian Guard looted the imperial palace, fearing that their days, and their high pay, would soon be numbered, they discovered Caligula's uncle, Claudius, hiding behind the curtains. Despite the fact that Claudius was said to suffer from cerebral palsy, and his stammer and limp had seen some regard him as an idiot, the Guard realised he was the key to their survival. Proclaiming Claudius as emperor, the Praetorian Guard rounded up Chaerea, and any other known conspirators, and had them executed. But, while Chaerea's plot had failed, over the coming years the Guard continued to be involved in further scandals and conspiracies, most notably in AD 193.

According to Cassius Dio, after murdering Emperor Pertinax, the Praetorian Guard tried to cash in on the power vacuum by putting the Roman throne up for auction to the highest bidder. Eventually selling control of the empire to former consul, Didius Julianus, for the enormous sum of 25,000 Roman sesterces per man, a civil war erupted, known as the Year of the Five Emperors. Less than three months after buying the throne, Julianus was executed.

With scandals such as this, the structure of the spoilt, privileged and traitorous Praetorian Guard was permanently altered in the late second century AD, when Emperor Septimius Severus dismissed its members and began recruiting bodyguards directly from the legions. Still, their run as the guardians of the Roman throne didn't officially end until the fourth century, when, in AD 306, the Praetorians tried to play the role of kingmaker one last time.

Installing Maxentius as the western emperor in Rome set off a dizzying chain of civil wars and rival claims to the throne before Maxentius and his Praetorians were confronted by Emperor Constantine at the Battle of Milvian Bridge. While the Praetorians supposedly made a valiant last stand along the Tiber River, they were soundly defeated, and Maxentius killed. Convinced that the Praetorians could no longer be trusted, Constantine disbanded the unit once and for all, reassigned its members to the outskirts of the empire, and oversaw the destruction of their barracks at the Castra Praetoria. Not many wept at their demise. By this time, they had developed a reputation as privileged bullies, who felt they were above the law.

I must confess that I once came very close to being inadvertently involved in a coup myself. During my time serving in Oman, my friend and predecessor, Captain Tim Landon, had built a special relationship with the sultan's son, Qaboos bin Said, having attended Sandhurst together. However, due to a high number of attempted coups and murders, the sultan decided to keep Qaboos under house arrest for his own safety. He was, though, allowed to receive a weekly visit from Tim.

At the time, Oman was a very poor country, with most Omanis living in squalor. Oil had recently been discovered but very little of the wealth it provided trickled down. As such, the country had hardly moved on from medieval times. While there were vague plans for electricity, water and a few hospitals, nothing seemed to materialise. There were just three hospitals and a dozen schools in all of Oman. Moreover, when sitting with my men in the mess hall I was horrified to learn that Dhofaris could not legally leave the country to work abroad. If they were caught having done so, they were banished from their homeland for ever. All the while, the old sultan cut himself off from the world, happy to leave his people living in the Dark Ages. It is no wonder that so many of his citizens looked to the supposed allure of the Marxists if this was all he had to offer.

I certainly found myself conflicted at this. While I could see the abject poverty all around, I was also in Oman protecting the man who was said to be responsible for it, albeit on the orders of the British government. At one point, I was so dismayed I threatened to return home. However, when I discussed this with one of my men, whom I had christened Mohammed of the Beard, he said, 'It is said that Prince Qaboos will rule in a while and, with the oil that will soon bring money, he will, thanks be to God, give us all that the communists now promise, but without changing our religion. If you British leave before that can happen, then the communists will take over without a doubt. They will force us to leave Islam or they will kill us.' With this in mind, and remembering I was a British soldier who was employed to protect the sultan, I decided to stay.

I was further persuaded when I actually met the sultan a short time afterwards at his Salalah Palace. Small in stature, he had a gentle face and voice, and above his white beard he also had warm brown eyes that twinkled. He was far from the tyrant I had envisaged. In fact, he reminded me a little of Santa Claus in his looks and demeanour. Since I had already decided that my job was to fight, and possibly die, for him, I was pleased that I found myself instinctively liking him, despite all my concerns.

Yet, unbeknown to me, Tim Landon and the sultan's son were planning a coup. Just weeks after my time in Oman had come to an end, they struck. While the sultan rested, a group of ten Omanis entered the palace compound unopposed. After a shootout with the sultan's bodyguard, the sultan himself was wounded and forced to sign a letter of abdication in favour of his son, Qaboos. At this, the sultan was allowed to leave the palace and board an RAF plane to London, where he spent the rest of his days living in the Dorchester Hotel in exile. He died in 1972, having eventually made peace with his son.

In three short years, Qaboos saw off the communist threat and heaved Oman out of the Middle Ages. With oil income booming, he made a series of huge investments in infrastructure, which included digging water wells in villages, building sixty hospitals with free medical care for all, and enrolling 34,000 people, of both sexes, into education. Unsurprisingly, he became a very popular leader, and continues to be to this day. I had certainly been aware that the sultan's power in Oman was waning during my time there, but I am not entirely sure I would have wanted any involvement in the subsequent coup, so it's a good thing I left when I did. My job had been merely to help protect the country from the communists and that is what I had focused on. Thankfully, it seems the country has thrived since I left and the coup was a force for good. I am now proud of the part that my men played, along with my platoon, in seeing off the Marxist threat, so that, in time, Qaboos could take over.

In ancient Rome, the life of the emperor continued to be under threat, with the demise of the Praetorian Guard. Soon they began to look elsewhere for protection, and, incredibly, one emperor turned to a bloodthirsty band of mercenaries from Norway, led by an exiled king ...

6

THE VARANGIAN GUARD

AD 988

Since AD 700, the Vikings of Scandinavia had become notorious in northern and western Europe for chasing wealth, along with displaying an insatiable appetite for violence. With their thick beards, horned helmets and ring-mail armour, they struck fear into anyone who crossed their path. Known to take drugs before a raid in order to increase their ferocity, they carried with them a deadly array of weapons. This included a long axe, which resembled a meat cleaver, as well as the Dane axe, whose sharp blade could sever a man's head in a single blow. It was, however, their double-edged sword for which they were renowned. Unlike a fencer, a Viking did not thrust with his sword, but instead hacked it down, hoping to split his enemy's skull or cut off an arm or leg.

They terrorised Europe with a series of raids, yet it is a mystery as to how they found their way there. They had no magnetic compass to guide them, nor has evidence been uncovered that proved they used a crude form of sun compass to navigate their way through the long periods of fog and choppy waters. In the early 1980s, when I completed the first open-boat voyage through the Northwest Passage from Inuvik to Ellesmere Island, I had great trouble navigating because the proximity of the magnetic pole rendered my compass useless. Moreover, I could barely see the sun due to relentless sea fog. It was only after a sponsor helpfully invented an infrared, hand-held device that I could detect the approximate position of the sun, and that was merely within 15 degrees of accuracy. In any event, somehow the Vikings found their way to these shores and their raids made them very rich men.

However, while some Vikings raped and pillaged Europe with abandon, others chose to head east, to Russia in particular, where they could sell their warrior skills to the warlords of the region. The very best of these men became highly sought-after, and were offered huge incentives, becoming enormously wealthy in the process. In time, this group of men earned the name the 'Varangians', which many scholars believe translates as 'vow of fidelity'. While this name no doubt referenced their incredible loyalty to each other, it was also their loyalty to their master that made them so valuable. This was certainly true for Basil II, then ruler of the Byzantine Empire, who was regarded as the most powerful man in the world.

By this stage, the Roman Empire was divided. Rome was the capital of the western half, while Constantinople, named after the Roman emperor Constantine, was the cosmopolitan capital of the eastern half. This became known as Byzantium, and its empire stretched across Italy, Greece, the Balkans and Asia Minor in the east. But, by AD 988, Basil's rule had been threatened by a series of internal and external revolts, not least by the rebel Vardhas Phokas. Already betrayed by his elite Greek bodyguards, Basil desperately sought loyal warriors who could help defeat his enemies and secure his empire. Rather than look within the Byzantine Empire, he decided he needed to recruit foreigners who lacked any political allegiances.

Many of the best Varangians were in the service of the Grand Prince of Kiev, Vladimir the Great, who was descended from Swedish Vikings and had recently taken power thanks to his army of Varangian warriors. This had earned them quite a reputation and news of them had certainly reached Basil's ears. Looking to strike a deal, he agreed to sell Vladimir his sister, Anna, in exchange for 6,000 Varangians. It was clear that Basil meant business.

The Varangians subsequently entered the emperor's service as his elite protection unit, with their quarters situated at the Great Palace in Constantinople. According to Alf Henrikson in his book *History of Sweden*, they were immediately recognisable by their long hair, a red ruby shimmering from their left ear and ornamented dragons sewn on their chain-mail shirts. But, more than that, they were soon acknowledged to be the best guards money could buy in antiquity, particularly when in AD 989 they swiftly crushed the revolt led by Vardhas Phokas. Such was the ferocious nature of their assault that not only did Phokas suffer a deadly stroke, but as his troops turned and fled the Varangians gave chase and mercilessly hacked them to pieces.

With Phokas out of the way, Basil turned his attention to quashing all potential revolts in the Byzantine capital, Constantinople. Acting as Basil's secret police, the Varangians ruthlessly cracked down on plots against him. Trawling the cosmopolitan city's back streets and taverns, they dragged any supposed traitors to the notorious Noumera prison, where the Varangians revelled in making an example of them. Their favourite methods included cutting off their nose or ears, rubbing their faces in human excrement, blinding them with acid, or even going so far as to castrate them with shears.

With the streets of Constantinople kept safe, and traitors severely dealt with, Basil now looked to expand his empire. Incorporating some of the Varangian Guard into his main Byzantine army he divided them into companies of 500, with each company put under the command of a regular officer. It meant the Varangians, unlike typical Vikings, now had to learn discipline, and this made them very dangerous indeed.

However, that is not to say their untamed ferocity was completely neutered. While most Vikings were used to charging headfirst into battle, the emperor used his Varangians in a different way. Holding them in reserve, he utilised them as shock troops, only sending them charging into battle when the fighting reached boiling point. Banging their shields in unison they would unleash a bloodthirsty cry as they launched themselves at their foe with barbarous intensity.

Indeed, when the Byzantine forces faced a seemingly indestructible defence, they looked to the Varangians to blast it away with their infamous 'Boar Snout' manoeuvre. Much like the Spartan phalanx, they packed tightly together, interlocking their shields for protection, while placing their heaviest, most aggressive warriors front and centre. However, rather than use the phalanx in a defensive formation, the Varangians would then charge into their enemy. The power and momentum thus generated frequently saw them burst through the enemy's line and spread panic in their ranks, as they unleashed a frenzy of bloodshed.

The addition of the Varangian Guard to Basil's regular forces saw them achieve a string of famous victories in Syria, Georgia, Armenia, Bulgaria, Greece and southern Italy. Such endeavours, and loyalty, saw the Varangians richly rewarded for their work. Earning between £17,000 and £32,000 per year, they also received bonuses as well as a third of all battle spoils, which could be very lucrative indeed. However, they made the bulk of their fortune whenever their emperor died. Upon this, they had the right to raid the imperial treasury and take as much gold as they could carry, a procedure known in Old Norse as *polutasvarf* ('palace pillaging'). And there were even more perks. With food, weapons and uniforms also provided, all of which were the very best that money could buy, they didn't have to spend a penny of their own on such items.

This wealth and prestige allowed the Varangian Guard to live the good life, especially in a bustling metropolis such as

Constantinople. Home to an estimated 1 million people, hailing from all over the globe, the city was a far cry from the bitter cold of Scandinavia. Boasting street lighting, drainage and sanitation, it also had hospitals, palaces with treasures from all over the world, public baths, churches filled with sacred relics, libraries and luxury shops, such as the 'House of the Lanterns', where luxury silks were sold, and which was lit up at night. Such splendours and marvels, all in a warm climate, were the daily background for the Varangians as they guarded the emperor wherever he should go. With their pockets bursting, they frequently headed to the chariot racing or bought virgin slave girls in the market for their pleasure. But nothing matched their love for alcohol. Indeed, such was their penchant for overindulging in Greek wine that the locals called them the 'emperor's wineskins'. Their drunken behaviour not only caused outrage in the taverns of Constantinople but also occasionally caused problems at work, where inebriated guardsmen were reported to have assaulted their own emperor.

Young, virile, muscular and vulgar, the Varangians flaunted their wealth and behaved like football hooligans. Having little regard or respect for anyone but themselves, their behaviour often appalled locals, especially when one drunken Varangian chiselled his name, 'Halfdan', into a wall at the church at Hagia Sophia.

Yet such wealth and prestige saw the Varangian Guard become the envy of family and friends. The Norse sagas even speak of the riches one Varangian displayed on his return home:

Bolli rode from the ship with twelve men, and all his followers were dressed in scarlet, and rode on gilt saddles, and all were they a trusty band, though Bolli was peerless among them. He had on the clothes of fur, which the Garth-king had given him, he had over all a scarlet cape; and he had Footbiter girt on him, the hilt of which was dight with gold, and the grip woven with gold, he had a gilded helmet on his head, and a red shield on his flank, with a knight painted on it in gold. He had a dagger in his hand, as is the custom in foreign lands; and whenever they took quarters the women paid heed to nothing but gazing at Bolli and his grandeur, and that of his followers.

It is no exaggeration to say that the Varangian Guard were considered to be the Premier League footballers of their age. And soon every Viking wanted to follow in their footsteps. Such was the exodus from Scandinavia that a law was passed declaring that no one was allowed to inherit any money while they served in the Byzantine Empire. Yet this did nothing to stem the flow of eager young warriors desperate to make their mark. Among those eyeing up a place in the Varangian Guard was an exiled king who would take the unit to a whole new level.

Harald Hardrada was born into Norwegian royalty in AD 1015. Renowned for his enormous size, and huge personality, at just fifteen years of age he helped lead an army to head off an uprising against his brother, Olaf. But he could only watch on as his beloved brother was killed in front of him. This moment was subsequently ingrained in his memory for the rest of his life. While he had to flee, he promised to one day return from exile, take his vengeance, and claim his rightful place on the throne.

But for now Hardrada needed a friend. As such, he made his way to Kiev, where his brother's ally, Yaroslav I, sat on the throne. While Hardrada was assured of a friendly welcome, Yaroslav was immediately struck by his physique.

Snorri Sturluson, who wrote the biography, *King Harald's Saga*, described him as physically 'larger than other men and stronger', with some suggesting he was over 7ft tall. He also stood out thanks to his long fair hair and bristling beard, while one of his eyebrows was said to be somewhat higher than the other. His hands were apparently so large that he could hold a human head within them, while his big feet propelled him across the ground in a flash. Yet Hardrada wasn't just a brute like so many of his fellow Vikings. He had also mastered horse-riding, swimming, skiing, shooting, rowing and could even play the harp.

With Yaroslav badly in need of military leaders, he subsequently made Hardrada captain of his forces, where he excelled in battles against the Poles, as well as against rivals in Estonia and the Byzantines. Word quickly spread of this prodigious Viking warrior and, in 1034, an opportunity presented itself that, as had many Vikings before him, Hardrada found hard to resist.

The Varangian Guard had heard of this impressive young warrior and now looked to recruit him to their ranks. For Hardrada, this offered an opportunity to recoup his fortune, while also building an army that could see him regain his crown. However, like all those before him, no matter his talents, Hardrada still had to pay a substantial fee before he could become a Varangian. This was something most recruits did without hesitation, knowing the riches that awaited them once they had joined the ranks.

By this time, Basil II was dead and had been succeeded by his niece, Empress Zoe, and her husband, Michael IV. Once again, the rulers, and the Byzantine Empire, were under threat and Hardrada and his men were desperately required to help them re-establish their authority. Following his own experiences, this was something Hardrada tackled with relish, having particular disdain for those who sought to usurp the crown.

While he fought victoriously in campaigns against the Pechenegs, he also helped to push the Arabs out of Asia Minor, where, according to his poet Arnorsson, he participated in the capture of eighty Arab strongholds. However, it was in Sicily that Hardrada would truly prove he was the finest warrior of his age. The stories of this time have been extensively detailed by Snorri Sturluson, as well as in Greek, Latin, French and Arab sources. While some believe them to be nothing more than folk tales, with similar accounts also attributed to other famous warriors throughout history, they nevertheless serve as thrilling stories.

Sicily had once been a Byzantine stronghold before it was captured by the Arabs in AD 902. For anyone who wished to control shipping, and trade, in the western Mediterranean holding sway in Sicily was vital, the island being a great meeting point between the Islamic, Greek and Latin worlds. For Michael and Zoe, recapturing Sicily would be a major coup, especially with so many rivals looking to usurp them.

However, taking Sicily would not be easy. When Hardrada landed on the island, he found that the Arabs had withdrawn into their fortified towns and had closed the gates. Before the age of gunpowder and cannons, this made it very difficult for any invading army to breach the walls. As such, castles and fortified towns were rarely taken. To storm a town wall would leave the attacker vulnerable to being shot from above by arrows or spears or to having boulders or boiling water dropped onto their heads.

In most cases, all that could be done was to lay siege to the towns and wait for the inhabitants to starve, or to succumb to disease. But Hardrada did not have the time to do this to every fortified town in Sicily. He needed results, and quickly. Yet this time around the sheer strength and ferocity of his fellow Varangians would not suffice.

Taking in the fortified town before him, Hardrada knew that he couldn't go over the walls so he came up with an alternative plan – go under them. He ordered his men to start digging near to a river, far from the prying eyes of the town's lookouts, where they could deposit the dirt and avoid suspicion. While Hardrada knew he had the element of surprise, he took no chances. With the tunnel dug, he waited for the time when the Arabs would be at their most vulnerable.

Sitting down to enjoy their evening meal in the candlelit great banqueting hall, the Arabs were stunned when the limestone floor suddenly burst open. Before they could react, Hardrada led his men up from the floor and viciously hacked to death all those who dared to defy him. The stunned Arabs were no match for these bloodthirsty warriors and soon the town was in the hands of Hardrada and the Varangians. And this was just the start of a string of astonishing victories.

Faced with another fortification, his methods became more cunning. Hardrada knew that this time the inhabitants might be expecting him to go under the walls so he devised a plan that would see the Arabs open the gate and welcome him with open arms. Setting up camp outside the walls, he was aware that the town's lookouts and spies were watching his every move. Therefore, he purposely placed his tent further away from the main Byzantine camp, as if to indicate he was unwell. His soldiers and medical staff also made a great show of rushing in and out of his tent, seemingly to tend to him. The Arabs not only took the bait, but they also joyously spread the word that Hardrada was dead, and the siege was over. The first part of Hardrada's subterfuge had gone like clockwork.

A delegation of Varangians subsequently approached the city and made an astonishing request. They asked that Hardrada be given a suitable burial for a man of his standing and suggested the city's church for such an event. Incredibly, the Arabs agreed. Having this famous warrior's remains in their town was not only an honour, but also a signal to all other invaders that, if they could see off Harald Hardrada, then they could see off anyone.

On the day of the funeral, just as Hardrada had planned, the Arabs happily opened their gates and allowed the Varangians, and Hardrada's coffin, into the city. However, as the coffin was escorted through the gates by pallbearers, priests and mourning Varangians, they suddenly dropped it to the floor. At that crucial moment, Hardrada sprang his trap. The Varangians pulled out their weapons and the coffin jammed the gate open, allowing all of their hordes to storm the city. Taking no prisoners, it soon turned into a massacre, and yet another fortified town fell to Hardrada and his men.

Soon no fort was safe. Before long, the Arabs were all on guard, intent on frustrating Hardrada at every turn. If he was to have any more success, he had to come up with his most ingenious plot yet. Yet, as Hardrada laid siege to a well-supplied town, its soldiers watched his every move from their fortress wall. There appeared to be no way in.

For days, Hardrada racked his brains. Watching the town, taking in its inhabitants' routines, he was desperate for any glimmer of inspiration that could help him crack the wall. Suddenly, birds flying from the town and settling in the surrounding fields caught his attention. He watched them day after day and saw that they would go from the fields and back to their nests in the thatched roofs of the town's homes. This gave him an idea.

Ordering his men to catch as many of the birds as possible, they then attached small splinters of wood to their backs, which were smeared with tar and then set alight. Hardrada and his men then set them free and watched as the burning birds flew back over the town walls, to their nests, setting the thatched roofs alight and engulfing the town in an inferno. The city gates were soon opened, and the Arab inhabitants came pouring out, surrendering without a fight.

One by one more towns fell before the whole island was back in Byzantine hands. Sicily was not only a great triumph, but it proved that the Varangians were more than just axe-wielding thugs.

As the years passed, however, Hardrada's favour in the imperial court declined following the death of Michael IV in 1041. The new emperor, Michael V, subsequently had Hardrada arrested and imprisoned. There is some disagreement about the reason for this. The Norse sagas state that Hardrada was arrested for defrauding the emperor of his treasure, while the English historian, William of Malmesbury, contends it was for defiling a noblewoman. Saxo Grammaticus, on the other hand, claims he was imprisoned for murder.

The sources also disagree about what happened next. Most, however, claim that during a revolt against the emperor the Varangians helped Hardrada to escape from prison, whereby he soon had his vengeance. After blinding the emperor, Hardrada watched as he was dragged kicking and screaming from his palace to be exiled to a monastery where he would spend the rest of his days.

Having escaped prison, Hardrada set his eyes on returning to Norway. With his huge fortune and loyal guards, he finally reclaimed his Norwegian crown, thanks in part to paying a large sum to his brother's son, Magnus the Good, the then king of Norway. But Hardrada soon began to look for further conquests and, in 1066, he set his eyes on England. It would, however, be his last. At the famous Battle of Stamford Bridge, he was struck by an arrow and killed. This would prove to be the last time the Vikings tried to conquer another country, although they did land in America before Christopher Columbus, but failed to realise the magnitude of their discovery and soon left.

While the Varangians continued to guard their Byzantine emperors, by the eleventh century, their masters were becoming increasingly concerned at the Muslim expansion that seemingly threatened Europe's way of life, as well as Christianity itself. To defend Europe, and reassert Christianity, a Byzantine emperor helped to fund a crusade to the Middle East. This would in time establish one of history's most controversial elite units and usher in decades of bloodshed . . .

7

THE KNIGHTS TEMPLAR
AND HOSPITALLERS

AD 1073

In AD 1073, Christians living in Jerusalem were rocked when the Sunni Muslim Seljuk Turks seized power from the relatively tolerant Egyptian Fatimids. As Christians fled the holy city in a panic, the reverberations of this seismic event were soon felt throughout Europe. With the Byzantine Empire already at war with the Seljuks in Asia Minor and Syria, and with Muslim armies overrunning Spain, France and parts of Italy, there was a real fear that not only was the empire itself under threat, but so was Christianity.

Increasingly concerned at this turn of events, the Byzantine emperor, Alexius Comnenus, desperately appealed to the west for aid in order to fight back. In 1095, his call was finally answered, when Pope Urban II publicly declared that all efforts should be directed towards funding a crusade to aid eastern Christians and recover the holy lands. In order to persuade a sufficient number of soldiers and holy men to enlist, the Pope promised them a remission of their sins. The response by western Europeans was immediate, with knights and peasants, rich and poor, flocking to the cross.

However, the First Crusade didn't get off to an encouraging start. The first group to make their way east was an undisciplined horde of French and German peasants. As soon as they reached Constantinople, they were annihilated by the Turks.

It was then that members of my family played a prominent role. Undeterred by these events, my ancestors, Godfrey de Bouillon and his brother Baldwin de Boulogne, raised an army of 40,000 knights and foot soldiers, by mortgaging much of their property. Along with another relative, Eustace Fiennes, this main crusading force set off for the Holy Land in 1096. Arriving at Constantinople in the vanguard, Godfrey was the first Crusader general to take an oath to the Byzantine emperor.

Finally reaching Jerusalem on 7 June 1099, Godfrey and his forces found that it had by now fallen back into Fatimid hands. Nevertheless, the Fatimids treated the Christians just as cruelly as the Seljuks had before them and, in the eyes of the Crusaders, needed to be banished. Moreover, they had not travelled all this way only to leave without taking Jerusalem for themselves.

However, the city was highly fortified and the Fatimids had poisoned the wells outside it. If they were to succeed, they needed to strike hard and fast. Yet attempts to breach the walls by ladder were met with boiling oil, rocks and arrows. At this, two Genoese ships, which had brought supplies to the Crusaders, were quickly torn apart for their timber and three siege towers were built.

On the night of 13 July, the Christians, led by my ancestor Godfrey, began fighting their way across Jerusalem's walls. As the Gate of Saint Stephen was forced open, the rest of the knights and soldiers poured inside and the city was captured. Tens of thousands of Jews and Muslims alike were swiftly slaughtered by the Christians, with chronicler Raymond of Aguilers recalling:

Piles of heads, hands and feet were to be seen in the streets of the city. It was necessary to pick one's way over the bodies of men and horses. But these were small matters compared to what happened at the Temple of Solomon, a place where religious services are ordinarily chanted. What happened there? If I tell the truth, it will exceed your powers of belief. So let it suffice to say this much, at least, that, in the Temple and porch of Solomon, men rode in blood up to their knees and bridle reins. Indeed, it was a just and splendid judgement of God that this place should be filled with the blood of the unbelievers, since it had suffered so long from their blasphemies.

The Crusaders had achieved their aim – Jerusalem was finally in Christian hands, with my relative, Godfrey, becoming the first 'Christian King of the crusader Kingdom of Jerusalem'. Heralded as a legend in Crusader circles, Godfrey was soon depicted in Tasso's epic poem *Gerusalemme Liberata*, Dante's *The Divine Comedy*, as well as in Handel's opera *Rinaldo*. Hailing from the Belgian side of the French border, a statue of him was erected in the Royal Square of Brussels in 1848, while he was voted by the Belgian public as the seventeenth greatest Belgian of all time in 2005.

But, despite the success of the First Crusade, violence in the Holy Land continued. In particular, groups of Christian pilgrims from across western Europe were routinely robbed and killed as they crossed through Muslim-controlled territories. Mounting corpses were soon left to rot along the route to Jerusalem, with the nadir for these attacks occurring around Easter 1119, when over 300 pilgrims were massacred.

During my time in Oman, I worked alongside many Muslim men. As a Christian, I was a real rarity but most treated me with curiosity rather than with any animosity. However, I remember on one occasion a young Kolbani, who had a permanent scowl, said to me, 'As a Christian, aren't you afraid of death, knowing you will burn in hell?' I had actually been waiting for someone to say such a thing so had already prepared a stock answer: 'We Christians are, like all Muslims and Jews, people of the Book, which clearly states that we will *all* go to Paradise. It is correct that the followers of the Prophet will go there before those who follow Christ, or those who killed Christ, but the Book does not say that we will spend the waiting period in Hell ... *Insha'Allah.*'

At this, there was much nodding of heads from those around me, although the Kolbani looked nonplussed and unconvinced. It remains a source of great mystery and sadness to me that there continue to be religious wars throughout the world. Is it not enough that we all believe in a god?

Yet, in Crusader times, as of now, this was clearly not enough, and the protection of pilgrims quickly became a priority. Just a few short months later, a French knight by the name of Hugues de Payens, along with eight relatives and acquaintances, came up with a solution. Founding the Poor Fellow-Soldiers of Christ and of the Temple of Solomon – later to be known simply as the Knights Templar – their goal was to protect Christian visitors to Jerusalem, as well as the city's holy shrines. With the support of Baldwin II, the then ruler of Jerusalem, the Templars set up headquarters at the city's sacred Temple Mount, from which they took their name.

The cream of Europe's knights soon flocked to the order, wanting not only to help defend their religion, but also to be absolved of their sins. Many criminals offered their services too, as they hoped it would rehabilitate them in the eyes of their families and God. In some cases, rather than send condemned knights to jail, law courts in the west sent them to fight in defence of Jerusalem.

Although the Templars were playing a crucial role in maintaining order in and around Jerusalem they were, however, initially poorly funded, with the knights having to wear donated clothes and even to share horses. However, in 1139, Pope Innocent II issued a papal bull that established the Templars as an independent and permanent order within the Catholic Church, who were answerable to no one but the Pope. This saw the Pope grant the Templars with special rights that certainly helped address their funding issues. These included being exempt from paying taxes, while all spoils of battle were to be theirs by right.

With this new influx of cash, the Templars set up a prosperous network of Europe's first banks, which allowed religious pilgrims to deposit assets in their home countries and withdraw funds in the Holy Land, so as to avoid the terrible consequence of being robbed of all they owned while travelling. As they began trading in the likes of wool, timber, olive oil and even slaves, many high-standing nobles also decided to leave their estates to the Templars on their deaths. In 1143, they inherited six castles from King Alfonso of Spain, as well as a tenth of royal revenues and a fifth of any lands conquered from the Muslims.

All of this made the Templars fantastically wealthy, and consequently very powerful. William of Tyre wrote of the order, 'They are said to have immense possessions both here and overseas, so that there is not a province in the Christian world which has not bestowed upon the aforesaid brothers a portion of its goods. It is said that their wealth is equal to the treasures of kings.'

This sudden accumulation of wealth and power saw conspiracy theories flourish. Some believed that upon undertaking excavations under Temple Mount they uncovered a vast fortune, a great secret, or even a sacred holy relic, such as the Holy Grail, which they then used to blackmail the church into providing them with special rights. All of this is unfounded but has made for great stories for the likes of novelist Dan Brown. But what isn't in doubt is that the Templars' fantastic wealth saw them build a series of castles, while also exerting significant influence.

While the Templars' initial mission was to defend and attack, there was another Christian order from around this time that had a very different goal. In 1080, a group of Italian merchants had established an order at the Hospital of Saint John in Jerusalem as part of a widespread charitable movement to help pilgrims, as well as the ill and the poor. In time, the order would be known as the 'Hospitallers'.

In 1113, in recognition of all its good deeds, the organisation was officially designated as a religious order by Pope Paschal II. In the same year, its first master, the Blessed Gerard, was officially appointed and its members recognised as monks. However, from 1120 the order was reorganised and made more militaristic by the then master, Raymond du Puy. Combining tending to the sick with defending the Crusader kingdom, the Hospitallers became one of the most formidable military orders in the Holy Land.

Despite being distinct orders, the Templars and the Hospitallers shared a number of beliefs and traditions. Both adopted a fraternal rule that emphasised the ideals of poverty, chastity and humility, but most important of all: obedience. Meals in both orders were to be eaten in complete silence, while ornate decor in clothing or battle raiment was forbidden. Any activity outside a martial or religious context was also seen as frivolous and inappropriate, thus chess, the raising of hounds, and hunting, were all frowned upon. Women were also said to be 'perilous things', who should be avoided at all costs.

The knights in both orders also looked alike. With their hair kept closely shorn, they sported a full beard to demonstrate humility, along with a simple brown or white habit for personal dress. However, in battle, the Templars wore a white surcoat, emblazoned with a red cross, while the Hospitallers began with a black tunic but eventually adopted a scarlet supravest, with a white eight-armed cross to be worn over their armour.

Personal property, beyond their necessary military equipment, was also severely frowned upon. Be it a small donation, or the pledge of an entire estate, a gift to a Templar or Hospitaller became a gift to the order; thus, while the brethren lived in relative austerity, the orders became increasingly wealthy.

While the brethren might have worn shabby habits in cloister, they possessed the finest equipment and weapons of war available in their day. Trained and disciplined to a degree almost unknown since classical antiquity, the military orders soon became the fire brigades of the Christian states, thrown into the most difficult actions and almost invariably outnumbered. The hospital rule even forbade a knight to retreat unless the odds against him were more than three to one. A chronicler from the period also stated that with the Templars in pursuit of the enemy they did not ask, 'How many are there?' but simply, 'Where are they?'

With the Templars and Hospitallers established, a Second Crusade was called for in 1147, following the threat of Zengi, a Seljuk Turk. While it ultimately ended in failure, with the Christian forces failing to capture Damascus, the rise of Saladin soon brought a greater threat to the Christian world.

Hailing from Tikrit, the same town that would in time spawn the Iraqi tyrant Saddam Hussein, Saladin was a fanatical Sunni Muslim who grew to become a trusted subordinate of the Syrian-northern Mesopotamian military leader, Nur al-Din. Participating in three campaigns into Egypt, which was governed by the Shiite Fatimid dynasty, Saladin became the last Fatimid vizier before changing the faith of the country from Shia to Sunni Islam in 1171. As you might imagine, this caused some consternation and led to him facing attacks from fellow Muslims, more on which later.

Saladin soon brought Aleppo, Damascus, Mosul and other cities under his control, which became known as the Ayyubid dynasty. Upon conquering each land, he also incorporated their armed forces into his own. His forces were soon formidable, consisting of Kurds, Arabs, Turks, Armenians and Sudanese, and at its centre was a corps of professional cavalry, trained and equipped not only in horse archery but also for close combat. While Saladin would command them from horseback, he would always be surrounded by his elite guard of Turkish Mamluks (slave soldiers).

Yet, to truly complete his empire, both religiously and politically, Saladin needed to capture Jerusalem. As such, with Saladin on the march, the Holy Wars reached boiling point on 4 July 1187.

With Saladin laying siege to the town of Tiberias, the Christian forces, led by Guy of Lusignan, king of Jerusalem, abandoned their defensive positions in Galilee and marched across the barren hills, in the blazing sun. When Saladin learnt that the Christians were on the march, he sent a series of skirmishers to harass and weary the vanguard and the rearguard, while his main Ayyubid force marched to Hattin, a well-watered village, from where they blocked off all roads to Tiberias.

After days of marching, and coming under attack from Muslim archers, the Christians were not only tired and thirsty but now they were trapped, with no access to water. On this waterless plain, the Christians had no choice but to spend the night stranded near the village of Meskenah, as the Muslims prepared for their final assault. Penning in the Christians, they spent all night singing, beating drums and chanting to strike terror into them. They also lit fires to make the Crusaders' throats drier and their thirst more extreme. Saladin relished having his enemy just where he wanted them, knowing that not even the knights of the Templars or Hospitallers could escape.

As the sun rose, Saladin blinded the Crusaders with the smoke from his fires, then ordered his archers to shower the enemy with arrows. With the Crusaders desperately trying to protect themselves, Saladin then unleashed his infantry to charge into them. His secretary, Imad ad-Din, has described the fearsome sight that now confronted the Crusaders:

A swelling ocean of whinnying chargers, swords and cuirasses, iron-tipped lances like stars, crescent swords, Yemenite blades, yellow banners, standards red as anemones and coats of mail glittering like pools, swords polished white as streams of water, feathered bows blue as birds, helmets gleaming over slim cur-vetting chargers.

The Crusaders were soon in total disarray. While some formed battle lines, others tried to flee, desperate to find water. Most were quickly shot down or taken prisoner. Some of those who did stay to fight, such as Count Raymond, managed to get through the lines and escape to the water supply at Lake Tiberias. However, for most there was no respite. The battlefield was soon littered with 'the limbs of the fallen, naked on the field, scattered in pieces, lacerated and disjointed, dismembered, eyes gouged out, stomachs disembowelled, bodies cut in half'.

Out of the 600 Templar and Hospitaller knights who fought in the battle, over 500 were killed, with Saladin decapitating all those he took prisoner. Imad ad-Din wrote of this:

Saladin ordered that they should be beheaded, choosing to have them dead rather than in prison. With him was a whole band of scholars and sufis and a certain number of devout men and ascetics, each begged to be allowed to kill one of them, and drew his sword and rolled back his sleeve. Saladin, his face joyful, was sitting on his dais, the unbelievers showed black despair.

Guy of Lusignan, however, could count himself lucky. Due to his status as king, not only was he spared, but he was released by Saladin just a year later after he swore to leave Palestine and never take up arms against Saladin again. This was, of course, a promise that would not be kept.

With the defeat at Hattin, and the annihilation of the most elite Christian units, Saladin swiftly took fifty-two Christian towns and fortifications, including Acre, Nablus, Jaffa, Toron, Sidon, Beirut and Ascalon. Jerusalem also fell on 2 October, with few forces left to defend it. The Muslims soon descended on Temple Mount, echoing cries of '*Allahu Akbar*', with any sign of Templar markings destroyed and returned to their earlier Islamic character. Those Christians who could afford to pay a ransom were allowed to leave, while those who were too poor were instead sold into slavery.

While news of the calamitous defeat at Hattin caused Pope Urban III to die of shock, the cry soon went up for a new crusade to the Holy Land. The elderly Roman emperor Frederick I Barbarossa responded immediately, setting out on 11 May 1189 with an army of 12,000–15,000 men, including 4,000 knights. However, while crossing the Saleph River on 10 June 1190, Frederick's horse slipped, throwing him against the rocks and causing him to drown. After this, much of his army returned to Germany.

Unsurprisingly, in 1189, King Guy resumed battle with Saladin upon his release. Along with his fierce rival, Conrad of Montferrat, he looked to retake the port town of Acre, which was one of Saladin's main garrison nodes and arms depots. The force defending Acre was significant, consisting of several thousand troops. In contrast, Guy's army was less than half its size. Despite numerous attempts to breach the walls, Saladin managed to hold the port, which eventually led to a fifteen-month double siege. With the Muslims in Acre entrapped, the Christians outside the walls were encircled by Saladin's forces.

Life both in the city and in the Christian camp quickly became intolerable. Food remained limited, while the water supply became contaminated with human and animal corpses, which led to the death of Guy's wife, Queen Sibylla, as well as their two daughters. As epidemics spread through the camp, and prostitution became rife, it seemed the Siege of Acre, along with the Third Crusade, was doomed to failure unless something remarkable occurred. Thankfully for the Christians, help would soon be arriving.

In 1189, Richard I, also known as Richard the Lionheart because of his reputation as a great military leader and warrior, joined forces with Philip II of France and together they finally defeated King Henry II, Richard's father, at Ballans. Just two days later, the king died, leaving Richard to succeed him. After Richard became king of England, he and Philip agreed to go on the Third Crusade, since each feared that during his absence the other might usurp his territories. However, there remained the issue of raising and funding an army.

Richard tried to raise and equip a new army by any means possible, even declaring, 'I would have sold London if I could find a buyer.' As such, he spent most of his father's treasury, raised taxes, and even agreed to free King William I of Scotland from his oath of subservience in exchange for 10,000 marks. To raise still more revenue, he sold the right to hold official positions, lands and other privileges to those interested in them. Those already appointed were forced to pay huge sums to retain their posts. However, in order to stand any chance of defeating Saladin, Richard knew he would require the services of Christianity's most elite forces. By this stage, the Hospitallers were in the process of being rebuilt under Fra' Garnier de Nablus and the Templars under Robert de Sable. While both orders had suffered significant losses at Hattin, they were eager to rejoin the battlefield and seek their revenge.

Recruiting the orders into their ranks, King Richard and Philip II set out on the Third Crusade in the summer of 1190. Joining Richard's side were more relatives of mine, Sir Ingelram Fiennes, Tougebrand Fiennes and John Fiennes. Sadly, all would be killed during the course of the crusade, with John's family giving the land in England where they buried his heart to the citizens of London; a place now known as Finsbury Square.

On 8 June 1191, Richard's troops finally landed at Acre, a few weeks after King Philip, with both aiming to end the siege that had commenced in 1189.

On the plague-infested battlefield, Richard, as always dressed in scarlet and brandishing his Excalibur sword and crossbow, led the cream of Europe's Hospitaller and Templar knights into battle. Despite being of similar numbers, they were far too strong for Saladin's ailing forces. After weeks of bombarding and assaulting the walls, Acre could hold out no longer, its inhabitants surrendering on 11 July. The two-year siege was finally at an end.

While Saladin had lost Acre, as well as his navy, which was stationed in the harbour, it was the massacre of thousands of prisoners, as told by Baha ad-Din, that really upset him:

> The enemy then brought out the Muslim prisoners ... about 3,000 bound in ropes. Then as one man they charged them and with stabbings and blows with the sword they slew them in cold blood.

Having successfully broken the Siege of Acre, Crusader forces now looked to capture the port of Jaffa before turning inland to reclaim Jerusalem. By this stage, King Philip had returned home, although he had left most of his forces under the charge of King Richard, which sources estimate numbered between 20,000 and 30,000 men. With the shocking Crusader defeat at Hattin in mind, Richard took great care in planning the march to ensure that adequate supplies and water would be available to his men. To this end, the army kept to the coast, where the Crusader fleet could support its operations. In addition, the army only marched in the morning to avoid the midday heat. This was important, as the knights would have been covered head to toe in chain-mail armour, with conical helmets over their heads, making them unbearably hot in the sun.

Departing Acre, Richard kept his forces in a tight formation, with his infantry on the landward side, protecting his heavy cavalry and baggage train. Responding to the Crusaders' movements, Saladin began shadowing Richard's forces in an attempt to stop them reaching Jaffa. As Crusader armies had proven notoriously undisciplined in the past, he began a series of harassing raids on Richard's flanks, with the goal of breaking up their formation. This done, his cavalry could then sweep in for the kill.

Rising on 7 September, the Crusaders needed to cover a little over 6 miles to reach Arsuf, just north of Jaffa. Aware of Saladin's presence, Richard ordered his men to prepare for battle and resume their defensive marching formation. Moving out, the Templars were in the van, with additional knights in the centre and the Hospitallers bringing up the rear.

King Richard's companion on this expedition, Geoffrey de Vinisauf, has described what happened next:

On all sides, far as the eye could reach, from the sea-shore to the mountains, nought was to be seen but a forest of spears, above which waved banners and standards innumerable. The wild Bedouins, the children of the desert, mounted on their fleet Arab mares, coursed with the rapidity of lightning over the vast plain, and darkened the air with clouds of missiles. Furious and unrelenting, of a horrible aspect, with skins blacker than soot, they strove by rapid movement and continuous assaults to penetrate the well-ordered array of the Christian warriors. They advanced to the attack with horrible screams and bellowings, which, with the deafening noise of the trumpets, horns, cymbals, and brazen kettle-drums, produced a clamour that resounded through the plain, and would have drowned even the thunder of heaven.

Under strict orders to hold formation, and despite taking losses from these hit-and-run attacks, the Crusaders pressed on. Seeing that these initial efforts were not having the desired effect, Saladin sent his mounted Ayyubid troops dashing forward to attack the Hospitallers with javelins and arrows. Protected by spearmen, the Hospitaller crossbowmen returned fire and began exacting a steady toll on the enemy. This pattern held as the day progressed, with Richard resisting requests from his commanders to allow the knights to counterattack, preferring to allow Saladin's men to tire before striking. Nevertheless, these requests continued, particularly from the Hospitallers who were taking the brunt of the attacks and were becoming increasingly concerned about the number of horses they were losing.

As the lead elements of Richard's army neared Arsuf, his Hospitaller rear of crossbow- and spearmen were fighting as they marched backward, desperately trying to keep the Muslims at bay. With his formation weakening, Garnier de Nablus again requested permission to lead his knights out, but once more he was denied by Richard. Assessing the situation, Garnier realised he could hold them no more. Ignoring Richard's command, he charged forward with the Hospitallers, as well as additional mounted units, unaware that fortune was to be on their side. For, at that exact moment, just as the Hospitallers charged, Saladin's cavalry had dismounted in order to better aim their arrows. Before they could react, Garnier's men burst from the Crusader lines, overran their position, and began driving back the Ayyubid right.

Richard might have been angry with the move, but nonetheless the Hospitallers had forced Saladin back, allowing the Crusaders to enter Arsuf and establish a defensive position. Regrouping his forces, Richard now ordered the Templars to attack the Ayyubid left and finish off the job.

Wielding their swords, the Templars not only succeeded in forcing the left to retreat but they were also able to defeat a counterattack by Saladin's Mamluk guard. With both Ayyubid flanks reeling, Richard personally led forward his remaining knights against Saladin's centre. This charge shattered the Ayyubid line and caused Saladin's army to flee the field into the wooded hills. However, with darkness approaching, Richard called off any pursuit of the defeated enemy, fearing that chasing them into the woodland would expose them to a trap.

Exact casualties for the Battle of Arsuf are not known, but it is estimated that Crusader forces lost around 700–1,000 men, while Saladin's army may have suffered as many as 7,000 casualties. This was a monumentally important victory for the Crusaders. It not only removed Saladin's air of invincibility, but it also provided the Templars and Hospitallers with revenge for Hattin.

However, such a victory did not mean Jerusalem could now be taken. Saladin found it easy to replace dead horses and dead soldiers in his own lands, while the Crusaders were far from home. As such, Saladin quickly recovered and focused on the defence of Jerusalem. When Richard marched to within sight of the city, he was advised by both the Templar and the Hospitaller grand masters that even if he took it, they would not be able to hold it without also controlling the surrounding hinterland.

With the two cancelling each other out, Richard and Saladin were forced to reach an agreement. Richard wrote to Saladin, 'The Muslims and the Franks are done for, the land is ruined at the hands of both sides. All we have to talk about is Jerusalem, the True Cross and these lands. Jerusalem is the centre of our worship, which we shall never renounce.' In response to this, Saladin wrote, 'Jerusalem is ours just as much as yours. Indeed, for us it is greater than it is for you, for it is where Our Prophet came on his Night Journey and the gathering place of the angels.'

After all the bluster, there was no getting away from the fact that a compromise would have to be reached. Eventually it was agreed that the Crusaders would demolish the walls of Ascalon, while Saladin would recognise the Christian positions along the coast. Free movement would also be allowed to Christians and Muslims across each other's territory, while Christian pilgrims would be permitted to visit Jerusalem and the other holy places. However, the Templars now had to make their new home at Acre, rather than return to Temple Mount. Nevertheless, the Third Crusade, in which Richard relied heavily on his elite knights, had saved the Holy Land for the Christians and kept the Muslim forces from invading Europe in greater numbers.

Richard would have a tumultuous return home, eventually dying from an infected wound sustained during a minor siege in France in 1199. European support of the military campaigns in the Holy Land also began to erode over the decades that followed his death, while the Templars and Hospitallers became fierce rivals. Their nadir came in the 1260s, when Templars and Hospitallers exchanged sword blows in the streets of Antioch. In the midst of battle against the Mamluks, a Hospitaller was said to have called out to a Templar, 'This night we shall have a word to say to each other in Paradise.'

'I doubt it,' replied the Templar, 'for my tent will surely be placed on the opposite side of that place from yours!'

By 1303, the Templar order had lost its foothold in the Muslim world and established a base of operations in Paris. There, King Philip IV of France resolved to bring down the order, perhaps because the Templars had denied the indebted ruler additional loans.

On 13 October 1307, scores of French Templars were arrested, including the order's grand master Jacques de Molay. Many of the knights were brutally tortured until they confessed to false charges, which included heresy, homosexuality, financial corruption, devil worship, fraud, spitting on the cross and more. A few years later, dozens of Templars were burned at the stake in Paris for their confessions. De Molay himself was executed in this fashion in 1314. Under pressure from King Philip IV, Pope Clement V reluctantly dissolved the Knights Templar, with the group's property and monetary assets being given to the Hospitallers.

It was the Hospitallers' charitable activities that saved them from the same grim fate that befell the Templars. When the Crusader principalities came to an end after the fall of Acre in 1291, the Hospitallers moved to Limassol in Cyprus and, in 1309, they acquired Rhodes, which they came to rule as an independent state. However, by the fifteenth century the Turks had caused them to flee and search for a new home. In 1530, the Holy Roman emperor Charles V gave the Hospitallers the Maltese archipelago, from where they saw off one of the most famous sieges in history, when the Turk, Suleiman the Magnificent, sought to invade in 1565. Eventually losing the island to Napoleon in 1798, they found a new home in Rome, where they gradually gave up warfare and turned wholly to territorial administration and medical care. The Hospitaller order still survives today, although its primary duties now involve issuing passports.

However, while the Middle East was dominated by the holy war between Muslims and Christians, there was one deadly group that did battle with both, not only targeting Saladin but also the future king of Jerusalem . . .

8

THE ASSASSINS

AD 1192

In the Crusader stronghold of Tyre, the extravagantly dressed Conrad of Montferrat emerged from the doorway of his friend, Philip, bishop of Beauvais, and happily walked down the dark, quiet streets. A key figure in the Third Crusade, Conrad had acted as chief negotiator in the surrender of Acre, and raised the king's banners in the city. And on this night, he had something to celebrate. Just days before, the barons of the Kingdom of Jerusalem had unanimously elected him as king, despite the protestations of King Richard and King Guy. After years of dispute and bloodshed, Conrad would finally be the most powerful man in the Holy Land.

Breathing in the fresh night breeze, Conrad walked down a narrow alley and happily thought of the future, when up ahead he suddenly saw two monks he knew well. Both were recent converts to the Christian faith, a source of pride for Conrad, as he hoped that, once he ruled, everyone in the region would shun Islam and do likewise. However, as he smiled at the monks, whom he believed to be his friends, he did not know that they were in fact agents of the Nizari Ismaili religious sect, otherwise known as the 'Assassins', and they'd been waiting for this exact moment for years.

The Assassins' story begins in AD 632 with the fracturing of the Islamic faith upon the death of the Prophet Muhammad. With a vacuum in the Islamic leadership, two forms of Islam soon developed, Sunnis and Shiites. In time, the Shiite branch grew popular with non-Arabs in newly conquered lands, especially Persia where converts developed their own traditions, customs and interpretations. However, in the eighth and ninth centuries a new Shiite faction emerged: the Ismaili.

While the Ismaili were initially a small, marginalised Shiite faction, the introduction of Hasan Sabbah to their faith would dramatically change their fortunes. Although Hasan's autobiography has since been destroyed, the historian Ata-Malik Juvayni read it in the thirteenth century and has provided much of the information we know today.

Born in 1052, in the town of Ray, just south of modern Tehran, Hasan was raised in the Shiite religion before converting to Ismailism in his teens. Obsessed, he soon commenced training to become an Ismaili missionary, first moving to Cairo and then returning to Persia, where he focused his efforts on the Elburz mountains along the southern shores of the Caspian Sea. Thanks to his charm and magnetic personality, he slowly but surely added more Ismaili converts to his ranks. Part of the lure was utilising one thing all potential converts had in common: an intense hatred of the Sunni Seljuks, led by Sultan Malikshah and his vizier, Nizam al-Mulk.

While it is likely that religion alone led to Hasan viewing al-Mulk as an adversary, some suggest that they had known each other since their schooldays, and that Hasan had even gone on to work for al-Mulk, taking on a post in his court. Apparently alarmed at Hasan's startling progress, it is said that al-Mulk deliberately discredited him in the eyes of the sultan, which resulted in Hasan leaving the court in humiliation. Whether or not this is true has been lost to time but we know for a fact that Hasan viewed al-Mulk as his enemy and sought to destroy him.

To achieve this goal, Hasan believed he needed a base for his converts, particularly with the Seljuks growing ever more hostile. In the late 1080s, he found just the place. Located in the rugged Elburz mountains, northwest of modern Tehran, the castle of Alamut sat high on a towering mountain, with just one track leading to it and deep gorges surrounding it. With its 360-degree views across the landscape, it was all but impregnable. Hasan knew that from here he could really build his faith, and his army. However, there was just one problem. The castle of Alamut was occupied by the Sunni Seljuks. If Hasan wanted it, then he had to somehow take it from them.

With limited military power, Hasan could not take it by force. But what he lacked in manpower he made up for in charisma and creativity. He subsequently sent his Ismaili agents into the surrounding communities to spread the faith before they were able to infiltrate the castle itself. Right under the noses of the Seljuks the Ismaili faith soon spread like wildfire. Before long, most of the castle garrison had been converted, and, without blood being spilt, Hasan entered the castle on 4 September 1090.

From this magnificent base, Hasan spread Ismaili propaganda and acquired more fiercely devoted followers who were unable to resist his magnetic personality. The twelfth-century Crusader, William of Tyre said of the Ismaili's extreme loyalty to their leader:

> The bond of submission and obedience that binds this people to their chief is so strong, that there is no task so arduous, difficult or dangerous that any one of them would not undertake to perform it with the greatest zeal.

Such was Hasan's devotion to his cause that he even sentenced his two sons to death for breaking his code. While one was accused of murdering a holy man, a charge later proved to be false, the other was executed for drinking wine. It was even said that he expelled one member of the sect for the crime of playing the flute.

As the Ismaili faith grew, so did its real estate, as Hasan seized, and built, more castles. Juvayni said of this, 'Wherever he found a suitable rock he built a castle upon it.' Soon the high valleys of Rudbar assumed the character of a miniature state – a heavily fortified Ismaili island in a Seljuk sea.

While the Seljuks did not initially respond to Hasan's taking of Alamut Castle, they could no longer ignore the spread of the Ismaili faith. Nizam al-Mulk described it as 'the pus of sedition' and vowed to eradicate Ismailism once and for all.

Hasan was well aware that, in a straight fight, his Ismailis could never hope to defeat the Seljuks' 300,000-strong army. But this led him to another idea. He could certainly cut off the head of the snake and let the main body wither and die. As such, he targeted Nizam al-Mulk for assassination.

To carry out such a mission, he looked to his most devoted followers – the *fida'iyin*. The *fida'iyin* were young, tough, resourceful mountain men, who were willing to follow their master wherever he might send them. Their recruitment initially involved an assessment of their physical fitness, fighting qualities and determined character, before Hasan would persuade them of his immense power and closeness to God. He did this by inviting them to a feast, where during conversation he would claim that he had the power to transport followers to Paradise. To achieve this, he would drug their food and drink. When they passed out, he had them taken to his secret, spectacular gardens, which he claimed to be Paradise.

I have seen similar brainwashing techniques during my time in Oman. In a school named after Lenin, 500 Dhofari children were forced to recite the thoughts of Chairman Mao parrot-fashion and to forget their Islam religion. Indeed, any mention of Islam was to be strictly punished. These young children were indoctrinated day after day, year after year, with Marxist propaganda. Any individualistic trait was stamped out and in time they became Marxist robots, trained to hate anything that did not follow this rigid doctrine. Once this had been cemented into their minds they were thrown into battle. It was heartbreaking to see these young men lose their lives for a cause they did not really believe in but had no choice but to fight for.

Many of the *fida'iyin* were no doubt the same, but with the supposed promise of Paradise, Hasan could persuade them to take on any mission. These men truly believed they were doing God's work and were not only prepared to lay down their lives for their religion, but were also expected to be rewarded in the afterlife for doing so. As such, it was almost seen as a reward for them, and their family, should they die when serving Hasan. When one *fida'i* returned home from a mission in which his compatriots had perished, his mother was so ashamed that her son had survived that she cut off her hair and blackened her face.

The fact that these chosen men happily embraced death is what truly made them terrifying. It is a mindset that has inspired countless terrorist groups over the years, such as Hamas, Hezbollah, al-Qaeda and ISIS. But there was one key difference. The *fida'iyin* refused to kill innocents. Instead, they struck specifically, cleanly and discriminately at their target.

Knowing he could rely on such men to do his bidding, Hasan picked out his elite squad and commenced their training. Over a series of months, nothing was left to chance. First, the chosen *fida'iyin* learnt how to gather intelligence on their target, with Hasan inserting an Ismaili spy into the Seljuk headquarters, who watched al-Mulk's every move.

To ensure success, Hasan demanded that the assassination be carried out at close quarters. An arrow or spear ran the risk of missing its mark, while there was always an antidote to poison. But there was another, more compelling, reason. He wanted the murder to shock the Seljuks by showing them that the Ismailis could get close to their greatest leaders. As such, he picked the *fida'i* he believed was best suited for this particular job, Bu Tahir, and trained him relentlessly.

Having received a blessing from his master, on 14 October 1092, Bu Tahir set out for Baghdad, where al-Mulk was currently residing. Intelligence had shown that al-Mulk could never refuse a blessing from a Sufi mystic. Bu Tahir entered al-Mulk's camp disguised in the robes of a holy man, while concealing a dagger beneath his clothing. Disappearing into the shadows of the night, he knew al-Mulk's routine like the back of his hand. First, al-Mulk would enjoy his nightly Ramadan feast. Then, with his guard down, he would be carried on his litter to his hareem. It was at this moment that Bu Tahir struck.

Approaching the vizier, with his head bowed, Tahir offered a religious petition for his consideration. Al-Mulk's guards instantly stood in his way, surrounding their master. But Tahir knew that al-Mulk would not be able to resist. Just as he expected, al-Mulk waved his guards away and beckoned the seemingly innocent mystic forward. It was a fatal mistake. Bu Tahir reached inside his robes, as if looking for the petition, but instead pulled out a knife. Before the guards could react, Tahir stabbed the old man in the chest. With his mission complete, Tahir merely accepted his fate and was killed by the guards, no doubt hoping he would now be rewarded in Paradise.

Al-Mulk's death was momentous news. For Hasan, it promised the freedom to really advance and maintain the Ismaili religion, as he said, 'The killing of this devil is the beginning of bliss.' The assassination, and its manner, also sent a powerful message to all of their heavily armed enemies – while they might be a tiny force, they could take on a vastly superior enemy, and win.

Over the coming years, the reputation of Hasan's Ismailis continued to grow. Appearing like ghosts, they would strike down their target with a dagger, then embrace death. The fact that so many of these kills were carried out in public, often before large crowds of horrified witnesses, ensured their reputation as cold-blooded killers spread far and wide.

While these kills achieved their objective, the Ismailis soon discovered that the psychological effect was far more powerful. Many of their enemies, and their followers, became too scared to go about their daily business, never sure when the ghostlike Assassins might strike. Soon anyone who thought they might be on an Ismaili hit list began to hire bodyguards and wear chain-mail armour under their clothing. This meant that the Assassins often didn't even need to kill their targets. The dread they inspired did the job for them. I saw similar tactics in operation in Dhofar. The *Adoo* would never be content with just shooting a man if a grislier, more shocking, death was available to them. Hence, they threw many men off cliffs in front of their fellow villagers. This had a profound psychological effect on them and soon brought everyone into line, thus ensuring the *Adoo* did not have to kill the whole village.

The Ismailis' tactics were so successful that no warlord or religious leader dared to attack them. However, their enemies soon learnt to fight back with their own unconventional means.

In an attempt to discredit Hasan and his followers, they spread wild stories, deriding them as crazy, brainwashed, drug-taking fanatics, who were manipulated by their evil leader high in the mountains. These notions were even repeated in *The Travels of Marco Polo,* which cemented the Ismailis' reputation as drug-fuelled thugs for centuries to come. Such accusations did, however, lead to the Ismailis becoming known by the name we all recognise today. The Sunnis apparently called Hasan and his band of Ismailis the 'Hashishi'. While this was supposed to reference their addiction to smoking hash, the nickname soon morphed into the 'Assassins'.

There is no evidence that Hasan or his Ismailis did actually take drugs of any kind. Modern historians believe that such was their devotion to their religion, and such was the precise nature of their kills, that it was very unlikely. In his book *The Assassins*, author Bernard Lewis believes that 'Hashishi' was a Sunni invention designed to denigrate Hasan and his followers. However, if anything, it appears it did quite the opposite. As the Assassins' legend grew, more followers flocked to the Ismaili cause, bewitched by its leader and the group's mystique.

While Hasan Sabbah died on 12 June 1124, the Ismaili Assassins fought on and flourished. Spreading as far as Syria the sect was later led by Rashid al-Din as-Sinan, who would become known as 'Old Man of the Mountains'. Sinan was similar to Hasan in a lot of respects; hypnotic, erudite, charming and highly intelligent, to many he was viewed as a god. Indeed, it was said that he would encourage this by using 'magical' tricks such as telepathy and clairvoyance.

His biographer, Kamal al-Din, describes him as 'an outstanding man, of secret devices, vast designs and great jugglery, with power to incite and mislead hearts, to hide secrets, outwit enemies and to use the vile and the foolish for his evil purposes'. The fact that he spoke little, and never ate in public, also gave him a superhuman quality in the eyes of his followers.

With a growing community of 60,000 followers, Sinan based himself in an impenetrable fortress known as Masyaf, where he watched with growing concern the Holy Land being invaded by Christian Crusaders. While this was a significant threat to the Ismaili way of life, it was the Sunni response to the Crusaders that seriously alarmed him.

As we have already seen, Saladin the Great was a fanatical Sunni who had changed the religion from Shiite to Sunni in Egypt, while he was also on the verge of uniting all the other Muslim states between Cairo and Aleppo. This would give him enormous power, which Sinan knew could seriously threaten Ismailism. As such, in 1175, Sinan placed Saladin as the number one target on his hit list.

With their methods now finely honed and tested, the Ismailis immediately went to work. Intelligence was always crucial and, as such, Sinan followed Hasan's lead and sent missionaries into Saladin's network, aiming to recruit more followers and glean from them what he could about his movements. But this proved to be frustrating. It appeared that Saladin was always surrounded by a close protection team. If they were to kill Saladin, they would need to take them out first. Still, Sinan gave the go-ahead for the mission to proceed.

Just weeks later, Sinan's men looked to strike as Saladin relaxed in his tent. But they were rumbled when a local emir raised the alarm. While the men tried to fight their way to Saladin's tent, they were soon struck down by his bodyguard. Saladin not only survived the attempt on his life, but was now aware the Ismailis were after him. This made the Ismailis' task ever more difficult, as Saladin now wore body armour twenty-four hours a day.

This forced Sinan to reassess his approach. While he realised that Saladin was prepared for his men, there was not yet any real urgency to kill him. He could afford to bide his time and wait for Saladin's guard to go down. To this end, he inserted a sleeper cell into Saladin's inner circle.

Almost two years passed before Sinan decided to strike again. With one of his agents now acting as Saladin's bodyguard, he gave the signal for the assassination to take place. However, it was again doomed to failure, with Saladin's body armour saving him.

Having survived two assassination attempts, Saladin decided not to take his chances and wait for a third. Marching with his huge army he lay siege to Sinan's castle at Masyaf and refused to leave until Sinan and his followers died of either starvation or disease.

There appeared to be no escape for Sinan. Outmanoeuvred, he knew that even killing Saladin would not be enough to extinguish the Sunni threat. All he could hope for was to somehow persuade Saladin to order his army away.

Kamal al-Din believes that Sinan, or one of his men, somehow infiltrated Saladin's camp and left a note in his tent, along with the Assassins' calling card – the dagger. The note apparently read: 'Death holds no fear for the *fida'iyin*. I will defeat you from within your own ranks.' This made it clear to Saladin that even if he killed Sinan and his men, they not only embraced death, but their followers would continue to hunt him down. What's more, the note proved that they could get close to him. Faced with this threat, Saladin agreed to speak to the Ismailis, which was just what Sinan had hoped for.

An emissary was subsequently sent from Masyaf to meet with Saladin in his tent, where the Sunni leader was surrounded by a fleet of bodyguards. At the meeting the emissary told Saladin he had a private message for him. At this Saladin sent out most of his entourage but still kept his two most trusted bodyguards by his side. Once again, the emissary told Saladin that the message was for his ears only. Saladin replied that he trusted his two body-guards like sons and whatever the emissary wanted to tell him he could also tell them. At this the emissary turned to the bodyguards and said, 'If I ordered you in the name of my master to kill this sultan, would you do so?' Suddenly they pulled out their blades and aimed them at Saladin's neck, saying, 'Command us as you wish.' Saladin was shocked. His most trusted bodyguards had been Assassins all this time, the very sleeper agents Sinan had planted in his organisation years before.

The emissary had made his point. If trusted men such as these could be Assassins, then Saladin could trust nobody. Saladin imme-diately lifted the siege and Sinan's Assassins were free, with Saladin knowing that should he ever strike back there could be Assassins lurking in his entourage ready to slit his throat. But the Assassins' triumph was short-lived. They still had to face the major force in the Middle East – the Christian Crusaders. With this, they eyed the assassination of the next king of Jerusalem, Conrad of Montferrat.

While religion was motivation enough for Sinan to kill Conrad, the actual reasons are unknown. Some claim that Conrad had captured an Assassin ship, killing the captain, imprisoning the crew and stripping the vessel of its treasure. When Sinan requested that the ship's crew and treasure be returned, he was rebuffed, and so a death sentence was issued. However, other reports claim it was a message to the Crusades, while some even believe it was a favour to Saladin, to whom Sinan had apparently grown close.

Whatever the reasons, by 1191, Conrad of Montferrat was a powerful and dynamic figure. In keeping with the tried and tested methods of the Assassins, Sinan inserted a sleeper cell into Tyre to watch his every move. His men learnt that Conrad enjoyed converting Muslims into Christians, so transformed themselves into Christian converts, taking on the identity of monks. Biding their time, they slowly ingratiated themselves with Conrad and his followers. But when news reached Sinan that Conrad had been elected the next Crusader king he could wait no longer.

As such, on 28 April 1192, as Conrad approached the two friendly monks in the alleyway he greeted them with a smile. Seconds later he was lying in a pool of blood as they stabbed him again and again. This was by far the most high-profile assassination of the Crusader age and its impact reverberated across the Christian world.

The poet Ambroise wrote of the assassination:

Two youths clad, who wore no cloaks, and each a dagger bore, Made straight for him, and with one bound, Smote him and bore him to the ground, And each one stabbed him with his blade, The wretches, who thus wise betrayed Him, were of the Assassin's men.

Within a year of his greatest triumph, Sinan was dead, yet the Ismaili faith, and his band of Assassins, continued to be the scourge of Sunnis, as well as Christian Crusaders.

There can be no doubt that the Nizari Ismailis were one of the most dangerous elite units the world has ever seen. Political assassinations have always been a prominent part of history, but the Assassins were one of the first special units to make it their modus operandi and they have become the poster children for terrorist groups all over the world, particularly through their unique brand of psychological warfare.

However, the Assassins were wiped out in the thirteenth century by an invading horde, who destroyed their castle at Alamut, and along with it many of their libraries and literature, which has made it difficult for historians to truly know their full story. Their conquerors were known as the Mongols, and, while religion and ambition appeared to have been the goal for conquests in Europe and the Middle East, it seemed the Mongols raped and pillaged countries just for the sheer pleasure of it ...

9

THE MONGOL KHESHIG

AD 1162

At its peak, the Mongol Empire was the largest contiguous empire in history, stretching from the Sea of Japan to the Mediterranean Sea and as far as the Carpathian mountains. With more than a million men enrolled in the armies of the khan, the Mongols became determined to conquer the world.

Unfortunately, just a single text of their exploits survives. *The Secret History of the Mongols* is therefore what most historians have used as their main source to learn about Genghis Khan, the notorious warlord and founder of the Mongol Empire, as well as his elite units that helped him conquer so many lands. While there is no way to prove the text's validity, it certainly provides for a thrilling story.

Born in north central Mongolia, around 1162, Genghis Khan was originally named 'Temujin', after a Tatar chieftain that his father, Yesukhei, had captured. Yesukhei was the leader of his tribe and it was expected that Temujin would one day follow in his footsteps. However, during a time of intense tribal warfare there were certainly no great lands or riches on offer. The vast steppe grassland was surrounded by impenetrable forests to the north and inhospitable desert to the south. And the steppe itself was unsuitable for agriculture, meaning the nomads had to rely on pasturing sheep and horses to survive.

As leader of his tribe, Yesukhei was a man to whom people flocked in order to ask favours or offer service. One such individual was a blacksmith called Jarchigudai. Clutching his infant son, Jelme, he had walked from the dark forest of the taiga, which was their home, and trudged through the snow, just to meet with him. When he found Yesukhei's camp, the blacksmith sought an audience with the chief, whereupon he offered him his first-born son as his servant. However, with his own son, Temujin, just born Yesukhei feared that his wife could not care properly for two infants. With gratitude, he sent the blacksmith away with the promise that, when Jelme had grown to be a man, Yesukhei would welcome him into his service.

The blacksmith returned to the forest with his son, and his wife soon gave birth to another boy, but the effort sadly killed her. The blacksmith called the boy Subotai and brought up his two sons alone. Meanwhile, the young Temujin was also soon to lose a parent, and consequently his status and inheritance, when his chieftain father was murdered by members of the rival Tatar tribe.

Upon hearing of his father's death, the 10-year-old Temujin tried to claim his position as clan chief but they refused to recognise the young boy's leadership and ostracised his family. Temujin, his brothers and his mother were subsequently abandoned on the steppe, without even so much as a horse between them. Over the next few years, they lived in abject poverty, with Temujin promising his mother he would one day have his revenge.

True to his word, Temujin rebuilt his family's fortunes. Displaying great bravery, tenacity and leadership skills, Temujin began to attract other men and their families to his entourage.

By the spring of 1187, he was established as the leader of a small group of followers, families and herds.

Not forgetting his promise to Temujin's slain father, the blacksmith travelled to meet him. *The Secret History of the Mongols* reveals their subsequent conversation:

'Many years ago, I had a son, Jelme, who was born when you were born and grew up when you grew up. When your people were camped at Deligun Hill on the Onan, when you, Temujin, were born, I gave your father a sable blanket to swaddle you in.' The old man could see from the expression on Temujin's face that it was the first time he had heard such a tale about his own youth. Every Mongol knew his lineage back at least five generations and could recite it at a moment's notice. But this, Jarchigudai sensed, Temujin had not known. The old blacksmith went on. 'When you were an infant, I also gave my son, Jelme, to your father, but since he was just an infant then I kept him with me.' He paused and looked at Jelme, who, he knew, was eager to join Temujin's clan. Since boyhood, Jelme had shown neither aptitude nor interest in becoming a blacksmith. Jarchigudai turned back to Temujin. 'Now,' he said, 'I have come to keep my promise to your father. Now Jelme is yours, to put on your saddle and open your door.' Then he gave Temujin his son.

While Jelme was to join Temujin, his younger brother Subotai was to return home and train to become a blacksmith like his father. Yet Subotai had other ideas. Inspired by his older brother's deeds, helping Temujin to gain victory in a number of battles, Subotai also offered his services just four years later.

However, when Subotai presented himself to Temujin he didn't appear to have much to offer. Unlike the Mongols, Subotai had not been taught to ride a horse by the age of three, nor was he given a bow, and trained in its use, by the age of five. He also had no experience of spending time in the saddle and navigating his way across the harsh open landscape, with few landmarks to guide him. It was skills such as these that had made the Mongol cavalry such a formidable force, whereby they could manoeuvre a galloping horse using only their legs, with their hands free to shoot arrows or stab and strike an enemy.

In the absence of such skills, it seemed all Subotai had to offer was being Jelme's younger brother. But this was good enough, as Jelme had proven to be a loyal and much-valued soldier. What Subotai lacked in experience he made up for in enthusiasm, fierce intelligence and unwavering loyalty during a time of warfare between the tribes, as seen in his pledge to Temujin, quoted in *The Secret History of the Mongols*:

> Then Subotai promised him: 'I'll be like a rat and gather up others. I'll be like a black crow and gather great flocks. Like the felt blanket that covers the horse, I'll gather up soldiers to cover you. Like the felt blanket that guards the tent from the wind, I'll assemble great armies to shelter your tent.'

Initially assigned to be Temujin's keeper of the tent door, the young Subotai received a first-class education in military planning. Taking in their every word, he listened to Temujin and his generals as they planned their conquests. He also threw himself into becoming a warrior by learning how to ride a horse and shoot a bow, and practising the manoeuvre-and-fire tactics of the Mongol cavalry.

Between 1197 and 1206, Subotai helped Temujin brutally overwhelm all of his enemies in bloodthirsty fashion, giving the Mongols a fearsome reputation. For instance, upon conquering the Tatar tribe that had killed his father, Temujin ordered all males to be led past the wheel of a wagon. All those who were taller than the wheel were beheaded, while the smaller children were incorporated into the Mongol army and the women forced into slavery. At this, the Tatars ceased to exist as a tribe. This was a tactic the Mongols would hone to perfection in time, ensuring that all conquered enemies would be rendered either extinct or broken as a people and as a country.

With Subotai rising to the rank of commander of several regiments, he helped Temujin finally destroy the Merkit tribe in 1206. For the first time in almost fifty years, all the Mongol clans were now united under the command of a single national leader. At this point, Temujin changed his name to one that would be known throughout history. He called himself Genghis Khan.

Having conquered his own land, Genghis immediately set about creating a national army capable of large-scale, sustained military operations against more powerful opponents. Within it, he made Subotai Lord of the Regiment and a field marshal. From that day forward, no major military operation was planned or undertaken by Genghis Khan, or later by his son, Ogedei, in which the voice of Subotai was not heard.

In addition to this, Subotai was enlisted into what many consider to be the Mongol army's most elite unit: the Kheshig. This was the home of the brightest and most promising of the Mongol army's military commanders and staff officers, of which Subotai was a leading figure. Its goal was to institutionalise excellence in command while also operating as an imperial guard.

The candidates for membership were identified early in their careers, with selection based strictly on merit. Therefore, unlike so many elite units of the age, common people were not excluded. All that mattered was that a candidate had shown great potential to be a great warrior or commander. Learning under the best the Mongol army had to offer, such as Subotai, candidates were groomed as officers and given a special status above the regular army.

This was reflected on the battlefield where the Kheshig imperial guard took its place next to the Great Khan in the centre of the line. In black lacquered armour, mounted on black horses, they were a terrifying sight. However, while protecting the king in battle was an important requirement, they also had another vital role: acquiring intelligence. And the lengths to which Subotai would go to acquire intelligence, and then use it to conquer lands, is one of the most extraordinary stories in military history.

While Genghis commenced his long war with China in 1211 (which would only come to an end in 1234), he brutally conquered Khwarizm in 1219, thanks in part to Subotai's strategic planning. The slaughter was so great that it was said that over four-fifths of the population were killed or reduced to slavery. As J. J. Saunders, in his book *The History of the Mongol Conquests*, puts it: 'The cold and deliberate genocide practiced by the Mongols ... has no parallel save that of the ancient Assyrians and modern Nazis.'

In 1220, as the Mongol army began its conquest of Afghanistan and Khorassan, Subotai had learnt of a land beyond the Caspian Sea where 'narrow faced men with light hair and blue eyes' dwelled. This gave him an idea. He proposed to Genghis that he, and another great Mongol warrior known as Jebe, along with their force of 20,000 men, be permitted to embark upon a long reconnaissance ride around these unknown lands, in order to provide intelligence for future conquests. Everywhere they travelled they would draw maps, take a census of the area, make surveys of the crops and yields, and even compile notes on the climate, to see if the area was worth conquering. They would also learn of the political and military situations, ascertaining how soldiers in these regions fought, while also working out who was a friend or foe, or who could be pitted one against the other.

Genghis approved of the idea, but ordered Subotai to return to Mongolia within three years so that these new conquests could then begin. As such, in February 1221, Subotai and his troop of

Mongol cavalrymen commenced a quite remarkable reconnaissance mission that has no equal.

Journeying into the unknown you might say is a speciality of mine, so I can somewhat understand how Subotai might have felt. Hundreds of miles from home, and unable to call for support, you have to ensure you are well prepared. Training for the terrain and conditions ahead is a prerequisite, as is preparing all the equipment you are likely to need. Raising the money for such journeys can also take time, as such expeditions are not cheap. For this reason, it took us seven years of training and preparation before my team and I were ready to embark on reaching both poles in 1979.

In 1992, when myself and Mike Stroud broke the record for the longest totally self-supported polar sledge journey ever made, so concerned were we at having everything we required, and like Subotai not being able to call for help, that our sledges weighed 485lbs. It was the equivalent of pulling a bathtub, with three average-sized adults in it, all through the snow and ice. I can only imagine how Subotai managed to handle the logistics of training and preparing 20,000 men for such a journey into what for them was the unknown. Ensuring every man, and horse, had sufficient calories for such a trek would have been a huge challenge. For the unsupported Antarctica journey, Mike and I ate over 5,200 calories a day and that still wasn't enough to sustain us. When Captain Scott and his men famously perished on their own attempt to reach the South Pole, they found 4,500 calories to be insufficient for such an endeavour and slowly starved to death. So concerned was I about this that, in the 1980s, I spoke to NASA about how they fed their astronauts in space. At the time, they were trying to invent lightweight food packs that contained a high number of calories. Unfortunately, on any of my subsequent expeditions, we were never able to get a daily ration of 5,000 calories weighing any less than those used by Captain Scott in

1910. But Subotai and his men would soon have more pressing matters to deal with.

After marching around the southern end of the Caspian Sea, along the banks of the Kura River and towards Tbilisi, they found the Georgian king, George the Brilliant, and 70,000 of his mounted knights waiting for them. This was supposed to be a reconnaissance mission, so Subotai had no intention of doing battle and losing valued men. However, when the Georgian army charged forward, he had no choice but to engage.

In any event, doing battle was a treat for these Mongol warriors. On horseback, they unleashed a barrage of deadly arrows, forcing the Georgians back, before they escaped into the night. While Subotai would not seek to engage the Georgians any further, he had seen first-hand how George the Brilliant's forces lined up, and the weapons they possessed. He stored the information away, as should they meet again, Subotai would be ready for them.

Yet Subotai was certainly not looking for further conflicts. If they had to fight pitched battles all the way across Georgia, his soldiers would be in no condition to make the passage across the Caucasus, Europe's highest and most difficult mountain range. This was not to say that the Mongols would not carry out raids on any soft targets they encountered along the way. After all, such conquests were in their DNA, while they also required a steady supply of horses and food, not to mention acquiring any easily available riches.

In Azerbaijan, they did just that: when Subotai's men captured the town of Maragha, after slaughtering the population they made off with as much booty as they were able to carry. Continuing on this looting spree towards Russia some towns merely opened their doors and gave Subotai whatever he requested, knowing full well the fate that would befall them if they did not comply. While the Mongols rightly developed a reputation for brutality, the mere fear they instilled saw many surrender without any need to do battle.

Subotai deliberately used such psychology to his advantage. While he might massacre one town, he knew that word of their savage brutality would soon travel to the surrounding areas. The next city would therefore be more likely to surrender without resistance, thus avoiding unnecessary Mongol casualties. The morality of this approach to warfare is no doubt open to discussion, but there can be no disputing that it was very effective. The Mongols' policy was pretty simple. Any city that surrendered without fighting would be spared. Those who caused the Mongol army to suffer casualties could expect no mercy.

In late autumn of 1221, when winter was already beginning in the foothills of the Caucasus, Subotai and the Mongol army advanced westward into Georgia, only to find King George once more waiting for them at the foothills of the Dagestan range. But, having seen the Georgian forces previously descend headlong into battle, Subotai was ready for them. This time, he ordered his cavalry to feign retreat and invited the Georgians into the foothills. Just as he had planned, the Georgians accepted the invitation and before they knew it they were totally outflanked. What followed was a massacre. The Georgian army – one of the finest in Christendom – was routed to such an extent that it ceased to exist, leaving Georgia completely without defence. For years afterwards, with no army to protect it, Georgia was continually ravaged by brigands and bandits.

With the Georgians vanquished, Subotai now had another issue to confront – crossing the deadly Caucasus mountains. Upon reaching the town of Derbend, Subotai stocked up on provisions and also accepted an offer from Rashid, the shah of Shirvan, to provide them with guides. But Rashid had no intention of allowing the Mongols to cross his lands.

Setting off in terrible conditions, the guides followed their master's instructions to take the Mongols over the worst possible route. Meanwhile, the shah sent his messengers the shortest way through the mountains to sound the alarm on the western steppes – the Mongols were coming!

The crossing of the Caucasus was disastrous. In blizzard conditions, the Mongols had to abandon their mangonels (catapults useful for breaking sieges), as well as much of their baggage, while hundreds of Mongol soldiers froze to death. During my expedition to Norway's Fabergstolsbre Glacier in 1970, we lost a sledge, which contained all of our skis and a couple of tents, when it slid out of control down a crevasse. We were lucky we did not lose more. My partners, Roger Chapman and Patrick Booth, were tied to it, and, if they had not slashed through the harness with a sawtooth sheath knife, they would have no doubt been killed. In such situations, you have no option but to thank your lucky stars it wasn't much worse and persevere onwards. From then on, the five of us on the team had to share a two-men tent, which at least made things warm and cosy in subzero temperatures.

Although, I must confess, I did wince at the thousands of pounds of gear we lost. Of course, if anything did go seriously wrong during my expeditions, I often had my wife Ginny on the other end of the radio, to send for help or supplies, which was always a tremendous comfort. Even if I did not particularly need help it was still wonderful to hear her voice, a luxury Subotai would not have had, being entirely cut off from his loved ones. However, on one occasion I could not even rely on Ginny for help. In 1982, during our Transglobe Expedition, Ginny's camp erupted in flames, which destroyed our precious stores. She could do nothing but watch as eight 45-gallon drums of stored gasoline exploded, soon followed by our rocket flares and ammunition. Newspaper and TV reports all over the world covered the story with headlines such as 'POLAR EXPEDITION IN FLAMES' and 'CONFLAGRATION AT POLAR BASE'.

There have been two particularly frightening situations where I have, however, had to call for help and be airlifted out. The first occurred when I attempted to become the first man to cross Antarctica solo and unsupported. After twenty-five days, and making excellent progress, I was stopped dead by kidney stones. While I attempted to ride out the most vicious pain I have ever experienced, I soon had no choice but to call for help, or risk doing permanent damage to my kidneys. The second occurred when I attempted to become the first man to reach the South Pole solo and unsupported. Unfortunately, a slab of ice moved, which opened up the sea beneath it, and took my sledge with it. Inside was seventy days' worth of food and thirty of fuel. To continue the expedition, I would need to rescue it, but to do so I had to quickly put my exposed left hand into the freezing water. While I managed to save the sledge, my fingers were riven by frostbite, and no attempt to warm them up would bring them back to life. Calling for assistance, I was soon airlifted out by skiplane. Sadly, the fingers were dead and beyond repair. Never one for patience, I eventually sawed off the dead ends myself in my shed, rather than wait for an operation.

With no luxury such as a plane to rescue them, the surviving Mongols finally forced their way through the pass, emerging exhausted and broken by the treacherous journey. However, there was to be no respite. Blocking their way from the narrow mountain pass was an army of 50,000 Cuman forces, and other assorted tribes.

For one of the few times in his long military career, Subotai found himself trapped. He knew he could not retreat back over the mountains, for a Muslim army raised by Rashid Shah would no doubt be waiting at the other end. With no means of retreat, and the narrow terrain depriving him of the ability to manoeuvre, Subotai ordered his exhausted army into a frontal attack, but they were soon repelled. With the situation looking desperate, Subotai realised he needed an urgent change of tack. Thankfully, he was a master at such things.

Sending emissaries to the Cuman tribe, he offered them bribes of gold and horses, while pointing out that the Cumans and Mongols were brothers of the steppes who had no reason to be at war with each other. The Cuman contingent took the bait. Riding away into the night, they left the remaining tribal contingents at the mercy of the Mongols, who promptly trounced them. But Mongol scouts also didn't intend to let the Cumans escape.

When the Cumans divided their army into two contingents, each going its separate way, Subotai and Jebe closed fast after the main body, caught up with it, and killed every last man. The treasure, and the valuable horses, which they had given as a bribe, were also recovered. With their enemies destroyed, the road to Russia now lay open.

As Subotai's forces made their way to Russia, they encountered merchants from Venice. Subotai immediately recognised them as a valuable source of intelligence about the west, and invited them into his camp. Lavishly entertaining them, he learnt all he could and had his men draw maps based on what the Venetians said lay beyond. As a result, provisional maps of Hungary, Poland, Silesia and Bohemia were designed and in time would all prove to be crucial. By the time the Venetians left, they had also signed a secret treaty with the Mongols in which they would send back detailed reports of the economic strength and military movements in the countries they visited. In return, the Mongols promised to destroy all other trading stations in the lands in which they rode, leaving Venice with a monopoly wherever Subotai's men went.

During the autumn and early winter of 1222, Mongol scouts and reconnaissance parties moved over the Don and Dnieper Rivers. Conducting forays into the Crimea they gathered information and reported on troop movements for future conquests. However, the Cuman forces that had survived the

Mongol slaughter now warned the Russian princes of their impending march.

On hearing this, the prince of Galicia, Mstislav the Daring, rallied the other princes to assemble their armies and meet the Mongol threat before it was upon their lands. Soon these forces numbered over 80,000 men.

Subotai learnt from his scouts of this advance and realised he had to move quickly to the east if he was not to be caught in the Russian multi-pronged advance. Once more Subotai attempted to use diplomacy to avoid conflict. He sent ambassadors to inform the Russians that they had nothing to fear from the Mongols, and, in any event, they were marching east, away from Russia. Unsurprisingly, the Russian princes failed to fall for this ruse, and continued marching onwards.

As Subotai's main forces moved east, he left a rearguard of 1,000 men to report on enemy movements, as well as to delay the Russians' approach. However, as the vanguard of the Russian army caught up with the Mongol rearguard, the sheer number of their forces soon overwhelmed them.

For nine days, the Russian army pursued the Mongols as they retreated north of the Sea of Azov. Yet, without any single commander, and riddled by indiscipline, the Russian contingents became strung out behind one another, over a distance of 50 miles. Meanwhile, the Mongols were riding over terrain they had previously reconnoitered; and Subotai used this knowledge to his advantage.

On 31 May 1223, the Mongols halted on the west bank of the Kalka River and formed for battle. As the Russian vanguard, under Mstislav, rode into the valley they were confronted by the Mongol army in battle formation. Without waiting for the rest of their contingent to catch up, the vanguard flew into action. This was just what Subotai was waiting for. With black smoke wafting across the battlefield from Mongol fire pits, the heavy cavalry suddenly appeared through the smoke and charged. The shock was tremendous. As some Russian forces retreated, others were slain, with the Mongols rampaging through their ranks. Chasing them for over 150 miles, they stormed the prince of Kiev's fortified camp, killing another 10,000 men, before sentencing the prince to death by suffocating him inside a box. At the end of the battle, a Mongol army of 18,000 men had slain over 40,000 Russians, including six princes and seventy nobles.

With their mission accomplished, and the three years coming to a close, Subotai, Jebe and their men made their way back to meet Genghis Khan on the banks of the Syr Darya River. But, before he departed, Subotai left behind scores of spies and secret messengers to provide regular reports on all that went on in Russia and eastern Europe. This information was to be passed to the Mongol intelligence service, who compiled dossiers on the various European countries and the political and religious rivalries that divided them. Subotai's reconnaissance into Russia was the longest cavalry ride in history, covering over 5,500 miles in just three years. The information he gained was to prove invaluable to the ceaseless growth of the Mongol Empire.

When Genghis Khan passed away in 1227, his son Ogedei succeeded him and aimed to continue with his father's plan to conquer Russia and Europe, with Subotai leading the way. Although his forces were stretched in campaigns in China, Korea and the Caucasus and the Black Sea, Ogedei provided over 150,000 men to Subotai for his mission, a number based on an estimate provided by intelligence officers. With all the information his spies had gathered over the years, Subotai finally set out in 1236 to conquer Russia and beyond.

Knowing that the key to conquering Russia was to attack each principality quickly, and in isolation, to prevent the formation of any coalition forces, the whole country was soon to fall to the Mongols in terrifying and bloody fashion. Nowhere was safe as fortresses were overwhelmed and turned into slaughterhouses. J. J. Saunders has written of this:

> Some were impaled, or had nails or splinters of wood driven under their finger nails. Priests were roasted alive, and nuns and maidens ravished in churches in front of their relatives.

The slaughter and destruction in Kiev was said to be so great that, when a traveller passed by the ruins of the city six years later, he described it as having only a few hundred huts, with the ground still littered with 'countless skulls and bones of dead men'. By the end of 1240, Russia was all but conquered and its forces destroyed.

With this, Subotai and his men now moved into Europe, utilising the invaluable information they had gained from the Venetians on their reconnaissance mission, knowing full well that many European countries' military forces were away fighting in the Crusades.

After a series of sweeping Mongol victories, the Hungarian king, Bela IV, desperately tried to block the Carpathian passes using felled trees, ditches, traps and other natural obstacles, in order to slow Subotai's forces while he tried to build his own. Despite these efforts, the Mongols still managed an astonishing pace of 60 miles a day, all while facing several feet of snow and obstructions blocking their path.

The combination of the winter invasion, the sheer speed of the Mongol advance, and Subotai's method of dividing his army into different divisions, saw the Mongols decimate the Polish and Hungarian armies in a series of battles. The Europeans not only had very limited intelligence on the invading Mongols, they were also no match for their firepower, mobility, training, command and endurance. This was emphatically seen at the Battle of Sajo River, where Mongol forces killed over 70,000 Hungarian soldiers – the entire royal army!

With the destruction of the Hungarian army, the Mongols controlled all of eastern Europe from the Dnieper to the Oder, and from the Baltic Sea to the Danube River. In four months, they had overwhelmed Christian armies totalling five times their own strength. By early 1242, Subotai even began discussing plans to invade the Holy Roman Empire itself. However, these plans were shattered when news arrived of the death of Ogedei Khan.

At this, the Mongol forces now had to make their way back home to take part in the election of a new khan. They were never to return to Europe.

Subotai continued to serve the khan and, following the campaign against the Chinese Song dynasty, he eventually returned to Mongolia in 1248. He subsequently spent the rest of his life at his home in the vicinity of the Tuul River, dying there at the age of seventy-two.

From a boy raised in the forests, he had risen from total obscurity to play a key role in establishing and maintaining the early Mongol Empire. As a member of the Kheshig, he had also passed on much of what he knew so that the empire could continue to conquer ever more enemies and lands, including that of the Ismaili Assassins.

In 1256, following an assassination attempt on Mongke Khan, the Mongols bombarded the Ismaili stronghold of Alamut. With Mongol warriors scaling the steepest escarpments, others positioned themselves on mountain peaks and used large Chinese-style siege crossbows to destroy the walls. This combination of force, firepower, and the eventual offer of mercy, worked. On 19 November 1256, the Ismaili imam surrendered to the Mongols, after which he was paraded from Ismaili castle to Ismaili castle, where he was made to order his followers to surrender.

However, after the death of Kublai Khan in 1294, the Mongols eventually disintegrated into competing entities and lost influence. This was in part due to the outbreak of the Black Death, but a group of slaves, who had become one of the most elite forces in the world, would also come into play ...

10

THE MAMLUKS

AD 1242

With the Mongol hordes rampaging their way across Europe, a tribe of Kipchak Turks tried to flee to safety. Crossing the Black Sea, they eventually made their way to Bulgaria, settling in a small village, praying they had finally found peace.

Yet life was still hard for the Kipchaks. Many were impoverished farmers and they found it almost impossible to grow crops in the cold Bulgarian winter. Some starved to death. But together they somehow found the will to go on, hoping to move on soon and reach warmer climes, where their crops could grow more freely.

However, on another bitterly cold night, as families retired to their makeshift homes, their peace was shattered by a blood-curdling cry. From the darkness, the Mongols suddenly charged through the village, dragging screaming families out of their homes and seizing anything of value. As one family after another was slaughtered, the Mongols hauled another Kipchak towards the executioner. But this Kipchak was different from the rest. Known as Baibars, this 19-year-old, blond-haired, blue-eyed young man, with a cataract in one eye, towered over the rest of his tribe. It was impossible to miss him.

Approaching Baibars with his blood-stained sword, the commander looked him over. In this moment, Baibars' life rested in his hands. Seconds later the commander turned and nodded to his inferiors; this man would be spared, he might be of value. However, everyone else in the village could die. Roughly placing Baibars in chains, he was taken away, but as he turned he saw a Mongol soldier stab his mother and father to death. It was a sight he would never forget, and an action the Mongols would live to regret. For this peasant slave would one day have his vengeance.

Taken to the slave market at Sivas, where only the very best men captured from across Europe were available for purchase, Baibars and other prisoners were put on display. Even in this company, Baibars stood out. It was clear to any buyer that, with his huge size, aggressive demeanour and ability to wield heavy weaponry, he would be an asset to any army, and at this time there was one army that specialised in turning slaves into fearsome warriors – the Mamluks.

The Mamluks came into existence in the ninth century as the Muslim caliphs sought to form a military force using enslaved men. As these slave soldiers were without regional, tribal or other personal ties, they believed they would be loyal only to the caliph. Like Baibars, they were mostly of Turkic origin, as from an early age such men possessed riding and archery skills and could be transformed into exceptional soldiers.

After a steady bidding war, Baibars was purchased by an Egyptian of high rank. However, having taken Baibars to Cairo, the Egyptian was arrested and his slaves confiscated by the sultan of Egypt, as-Salih Ayyub. With this, Baibars was not only to serve a new master, but was also to be trained in the ways of the Mamluks.

Taken to their barracks in the Citadel of Cairo, his every hour was soon devoted to military training, which included archery and mounted combat drills. Great emphasis was also placed upon the *Furūsiyya*, which was not dissimilar to the chivalric code of the Christian knight, insofar as it included a moral code embracing virtues such as courage, valour, magnanimity and generosity. It also addressed the management, training and care of the horses that carried the warrior into battle and included cavalry tactics, riding techniques, armour and mounted archery. Military tactics were also studied, including the formation of armies, and how to utilise fire and smokescreens. Such was the desire to create a fearsome fighting machine that even the treatment of wounds was addressed. In such company, Baibars was a star pupil, excelling in combat and military tactics.

While the Mamluks trained relentlessly, and lived almost entirely within their garrisons, some leisure activities were encouraged. Polo was particularly popular, as the need for control of the horse, tight turns and bursts of speed perfectly mimicked the skills required on the battlefield. Mounted archery competitions, horseback acrobatics and mounted combat shows, similar to European jousting, also took place up to twice a week, while a hippodrome in Cairo would eventually be built where Mamluk games took place.

It was just as well that Baibars and his fellow Mamluks were trained to such a high standard. At this time, the Christian Crusaders were looking to take Egypt and soon the sultan hurled his legions into combat. Here, Baibars made a name for himself. One Crusader described him as 'brilliant as Caesar, and as harsh as Nero'. Quickly rising through the ranks, he was soon transferred to the Bahriyya Regiment, a unit of 1,000 Mamluk shock troopers, who served as the elite bodyguard to the sultan.

Some believe that Baibars truly made his mark in the 1244 Battle of La Forbie. Earlier in the year Egypt's allies, the Khwarazmians had captured Jerusalem. Determined to reclaim it, Crusader forces marched towards La Forbie, a small village northeast of Gaza, where Ayyub's Mamluks were waiting. Despite the Christian knights' superior numbers, the Mamluks were able to destroy them, killing 5,000 Crusaders and taking 800 prisoners, many of whom were famous warriors and nobles.

Some have speculated that Baibars was in command of the Mamluk forces in this battle, but the historian Stephen Humphreys, in his book *From Saladin to the Mongols*, disagrees. He claims that there was a warrior with a similar name, which has led to this confusion. In any event, the Crusaders' humiliating defeat at the Battle of La Forbie saw Pope Innocent IV call for a Seventh Crusade. King Louis IX of France was one of the few to answer this call. By 1248, he had assembled a 15,000-strong army, which included 3,000 knights and 5,000 crossbowmen. After months of travel, thirty-six of Louis' ships landed at Damietta on the Nile. Meeting little resistance from the Egyptians and swiftly taking the port, Louis hoped to build a base from which he could attack Jerusalem. The fifteenth-century Muslim historian al-Maqrizi has recorded Louis IX subsequently sending a letter to as-Salih Ayyub that read:

As you know that I am the ruler of the Christian nation, I do know you are the ruler of the Muhammadan nation. The people of Al-Andalus give me money and gifts while we drive them like cattle. We kill their men and we make their women widows. We take the boys and the girls as prisoners and we make houses empty. I have told you enough and I have advised you to the end, so now, if you make the strongest oath to me and if you go to Christian priests and monks and if you carry kindles before my eyes as a sign of obeying the cross, all these will not persuade me from reaching you and killing you at your dearest spot on earth. If the land will be mine, then it is a gift to me. If the land will be yours and you defeat me, then you will have the upper hand. I have told you and I have warned you about my soldiers who obey me. They can fill open fields and mountains, their number like pebbles. They will be sent to you with swords of destruction.

Unsurprisingly, Ayyub refused to surrender, especially as his forces had been so successful against the last Crusaders at the Battle of La Forbie. However, as Louis marched towards Cairo, Ayyub suddenly died, leaving a huge power void. Victory for the Crusaders now seemed inevitable, especially when they quickly overwhelmed the Egyptian camp at Gideila, and then turned their attentions to the nearby walled city of Al-Mansurah.

As Egyptian forces fled from the city, Baibars and his Mamluks stood firm, despite facing impending defeat. Unless they devised a plan quickly, they would all be slaughtered and Egypt, as well as Jerusalem, would be lost.

Watching the Crusaders as they approached the city in hordes, Baibars suddenly took control and issued an order that surprised his fellow Mamluks. 'Open the gates,' he cried, knowing that they were in no shape to sustain any siege and this might be their only chance.

Moments later, the Crusaders rushed into the walled city, thinking that the Egyptians had abandoned it. As they rode on horseback through the empty streets, Baibars now ordered his men, who had been hiding in the shadows, to strike. With the gates slamming behind them, the Crusaders suddenly realised they had fallen into a trap. Unable to escape, and with their cavalry useless in the narrow streets, they were ambushed by the Mamluks, as well as by thousands of people from the town. Dragged from their horses, the Crusaders were overwhelmed and murdered where they fell. Such was the outpouring of un-restrained violence that only five Templar knights escaped alive, one of whom later lamented:

Rage and sorrow are seated in my heart ... so firmly that I scarce dare to stay alive. It seems that God wishes to support the Turks to our loss ... ah, lord God ... alas, the realm of the east has lost so much that it will never be able to rise up again. They will make a Mosque of Holy Mary's convent, and since the theft pleases her Son, who should weep at this, we are forced to comply as well ... Anyone who wishes to fight the Turks is mad, for Jesus Christ does not fight them anymore. They have conquered, they will conquer. For every day they drive us down, knowing that God, who was awake, sleeps now, and Muhammad waxes powerful.

From what had looked to be impending disaster, Egypt now stood victorious, and it was all thanks to a Mamluk slave. But any thanks were in short supply.

Following the sultan's death, a series of power struggles saw the throne eventually go to his widow, Shajar al-Durr, who subsequently married a Mamluk called Izz al-Din Aybak. However, al-Durr abdicated the throne shortly afterwards, leading to Aybak becoming the first Mamluk sultan in history. But this was not good news for Baibars, far from it. Aybak and Baibars were enemies, and should he stay in Egypt then his life would be in danger. With no other option, Baibars, together with Mamluks loyal to him, fled to Syria. In the meantime, the power struggles in Egypt continued, with Aybak being murdered by his wife in 1257 and al-Muẓaffar Sayf al-Dīn Quṭuz becoming the new sultan.

While Baibars and his Mamluks made a living in Syria as mercenaries, the new sultan urgently reached out to him in 1260. Not only did he invite Baibars to return to Egypt, but he also offered a chance to settle old scores: the Mongols were coming.

In February 1258, the Mongol armies of Hulegu, grandson of Genghis Khan, had taken Baghdad, where 250,000 were said to have been killed. Mongol troopers even rolled al-Musta'sim, the last Abbasid caliph and spiritual leader of Islam, into a carpet before kicking him to death. Damascus and Aleppo fell soon after, the Mongols completing their conquest of Syria with the near-annihilation of the Assassins. The only thing that now stood between the Mongol Empire and Jerusalem, Mecca and Cairo were the armies of the Egyptian sultan, and Baibars.

Returning to Cairo, Baibars immediately took command of the Egyptian army, planning not only to defend his adopted country but also to have his vengeance against the enemy who had murdered his family before his very eyes. On 3 September 1260, he would have his chance to avenge them. Marching to Ain Jalut, just north of Jerusalem, he found 30,000 Mongols waiting for him. The Egyptian forces were more than double that of the Mongols but, as we have seen previously, this was no guarantee of a Mamluk victory.

Firing their deadly composite bows into the Egyptians, the Mongols followed up with a charge of heavy cavalry that smashed the Egyptian lines and sent the entire army into a full retreat. As the Mongols cut the fleeing Mamluk warriors down, Baibars led them towards the surrounding forests, apparently in retreat. Yet, just as they approached the woodland, he let out a cry. Suddenly, another 60,000 Egyptian horsemen thundered out of the trees and slammed into the Mongol flanks, cutting off their only escape route. Remembering what they had once done to his village, Baibars ordered no survivors. The Mongols were butchered to the last man. It was the first defeat the horde had suffered, and Ain Jalut marked the furthest their empire would ever stretch in the Middle East.

Having achieved a feat most felt impossible, Baibars understandably believed that his deeds deserved reward. While he eyed the governorship of Aleppo, he was dismayed to find that no such gift would be forthcoming. Sultan Qutuz now feared Baibars' popularity, as well as his ambition, and wanted to see him banished once again. But, this time, Baibars was unwilling to go quietly. Assassinating Qutuz, Baibars became the sultan of Egypt and Syria, marking an astonishing rise from the young peasant who had been kidnapped and sold into slavery by the Mongols just sixteen years before.

Ruling an empire that stretched from Cairo to Baghdad for over seventeen years, the ex-slave was so beloved that he became known as the 'Lion of Egypt', particularly for his battles against the Crusaders. Attacking the last Crusader towns and villages in Syria and Palestine, he soon retook the likes of Arsuf and Jaffa, while he also devastated the Templars at their last stronghold of Antioch, as well as the Hospitallers at their Krak des Chevaliers castle.

He was, however, to die in mysterious circumstances in 1277. Some sources claim he died as a result of drinking poisoned kumis, which was intended for someone else, while others suggest that he may have succumbed to a wound sustained while campaigning. But, while the Crusaders believed him to be the personification of evil, to Muslims, he died a great hero, with his Mamluk descendants remaining in power until 1517, when their dynasty was eventually extinguished by the Ottomans, and their own elite force of slaves . . .

11

THE OTTOMAN
JANISSARIES

AD 1453

Since AD 324, Constantinople had acted as the capital of the Byzantine Empire, becoming the jewel of Christendom. From the Arabs to the Vikings, many had tried to take it, but all had failed. Defended from the sea by its large navy, its legendary Theodosian Walls, surrounded it on land, making it all but impregnable.

Extending across the peninsula for some 6.5 kilometres, the walls were 20 metres wide, and boasted a 7-metre-deep ditch, which could be flooded with water fed from pipes. Behind that was an outer wall, which had a patrol track to oversee the moat. Then came a second wall, which had towers, and an interior terrace, so as to provide a firing platform to shoot down on any enemy forces who were attacking the moat and first wall. Beyond that wall was a third inner wall, which was almost 5 metres thick, 12 metres high, and presented to the enemy ninety-six projecting towers. Each tower was placed around 70 metres distance from another, reaching a height of 20 metres, and could hold up to three artillery machines.

Unsurprisingly, Constantinople, was considered virtually unassailable. But, in 1453, it faced its greatest challenge yet, as a new empire conquered all before it, and looked to make Constantinople its own capital.

The Ottoman Empire had begun as a small Turkish emirate in the late thirteenth century. Within 100 years, it had expanded into Thrace, Thessaloniki and Serbia, and, after its army had defeated the Crusaders, it looked to storm Constantinople. Yet its initial attempts, in 1394 and 1422, were both unsuccessful. Determined to claim it, the rise of Sultan Mehmed II in 1444 saw the Ottomans look to finally sweep away the Byzantines with two secret weapons; gunpowder, and an elite military unit known as the Janissaries.

Much like the Mamluks the Janissaries were also slave warriors. Every five years Turkish administrators would scour their regions for the strongest boys aged between eight and twenty, who would then be taken from their parents to become Janissaries. Rather than be distressed, many Christian families were actually happy with this because it offered their sons the possibility of social advancement. Many Janissaries progressed to become colonels or even statesmen, opportunities that would not otherwise have been available. There were families who were of course upset to see their sons snatched away, and as it was forbidden for married men to become Janissaries, this saw many young boys married off at a young age.

Taken to Istanbul (as the Ottomans were to rename Constantinople), the boys were inspected before two lists were made. The most intelligent were sent as 'inner (service) boys' to the Sultan's Palace Schools, destined, with luck, for high office. There, they studied religion, as well as Turkish, Persian and Arabic literature. Physical activities were also encouraged, with the boys engaging in horse-riding, javelin throwing, archery, wrestling, weightlifting and even music. Great emphasis was also put on honesty, loyalty and good manners, traits that were seen as necessary for any budding statesman.

Those boys who didn't make the first list were known as the 'foreign boys'. These were sent to the households of senior or respected men for the first phase of their education, learning Turkish, and Islam, as well as the customs and cultures of Ottoman society. After five to seven years of this, the boys were then gathered for training at the Enderun *acemi oğlan* – 'cadet' – school in the capital city, where they spent the next six years training in different areas as engineers, artisans, riflemen, clerics, archers, artillery and so forth.

Living in monastic conditions, the Janissaries trained under strict discipline, much like the Knights Templar and Hospitallers. Expected to remain celibate, unlike other Muslims, they were also expressly forbidden to wear beards, although a moustache was tolerated. They were also encouraged to follow the dervish saint Haji Bektash Veli, who served as a kind of chaplain. As a symbol of their devotion to the order, they wore special hats called *börk*. These hats had a holding place in front, called the *kaşıklık*, for a spoon. This symbolised the so-called 'brotherhood of the spoon', which reflected a sense of comradeship among the Janissaries who ate, slept, fought and died together.

Protectors of the throne, the Janissaries were expert archers, who were also handy with the axe, club or dagger. However, it was the double-curved Turkish yataghan sword that was regarded as their signature weapon, which when slashed across an enemy could virtually cut them in two.

Making up a tenth of the overall Ottoman army, the Janissaries saw action in all major campaigns, although they enjoyed far better support than other armies of the time. Part of a well-organised military machine, one support corps prepared the roads, while others pitched tents and baked bread. Their weapons and ammunition were also transported and resupplied by the so-called Cebeci corps. Medical teams of Muslim and Jewish surgeons also joined their campaigns, with the sick and wounded evacuated to dedicated mobile hospitals set up behind the lines. In short, their every need was catered to so that they could concentrate on their one job – protecting the sultan at all costs.

In battle, during the strategic fake retreat of the Turkish cavalry, the Janissaries would hold the centre of the army against enemy attack. Yet it wasn't just on the battlefield where their skills were utilised. They were also adept at breaking sieges. Between teams of explosive experts, engineers and technicians, sharpshooters and sappers, they almost always prevailed, even against the most stringent defences. However, something that made the Janissaries truly stand out as the elite force of their age was the discovery of gunpowder by the Ottomans in the fourteenth century. While other empires had also put it to good use, it was the Ottoman Janissaries who adopted it faster, and far more extensively, than any of their rivals.

While gunpowder was invented in ninth-century China, Professor Kenneth Warren Chase, author of *Firearms: A Global History to 1700*, credits the Mongols with introducing it to Europe, with several sources mentioning Chinese firearms and gunpowder weapons being deployed against European forces at the Battle of Mohi in 1241. By the 1380s, the Ottomans were acquainted with gunpowder weapons, and before most others they integrated them into their standing forces. Furthermore, from the 1390s on, preceding their rivals by centuries, the Ottomans established a corps of permanent salaried troops who specialised in the manufacturing and handling of firearms. Indeed, from the fifteenth century, the Cebeci was created to look after and carry the infantry Janissaries' weapons. The army also had its own gun carriage drivers (*arabaci*) whose job was to manufacture, repair and operate war wagons in campaigns.

I recall the first time I held a gun. I was only very young and had found my father's loaded old service pistol hidden under my mother's pillow. One day, when my mother was out, I took the gun and threatened to shoot our cook, Christine, unless she gave me a slice of chocolate cake from the locked larder. She screamed so hard she scared me off, and told my mother of my misdemeanour, who quite rightly beat my hand with a cane.

In any event, by the early fifteenth century, the Janissaries' firepower overwhelmed that of most of their adversaries. However, while they were heavily armed, and well trained, in firearms, the Ottomans knew that even their skills might not be enough to breach the walls of Constantinople. To break the siege, they needed something that previous besiegers of Constantinople had lacked: cannons.

A Hungarian engineer named Urban had first offered his cannons to the Byzantines but Emperor Constantine could not meet his asking price. As such, Urban turned to Sultan Mehmed, who, well aware of the cannon's potential, offered him four times the price he had asked to ensure they were his. Soon Mehmed had a fearsome array of cannons that could destroy most castle walls; the largest measured 9 metres long and had a gaping mouth 1 metre across. Able to fire a ball weighing 500kg over 1.5 kilometres, it destroyed anything in its path. However, it could only be fired seven times a day, as it needed time to cool down between shots. Still, the Ottomans had plenty of smaller cannons, each capable of firing over 100 times a day, and each capable of causing extensive damage. This was a significant game-changer in siege warfare. No longer were the tricks of the likes of Harald Hardrada required. Now an army could blast their way through fortress walls.

Sadly, my own brush with explosives ended in ignominy. Having just been accepted into the SAS, I had managed to keep some of the detonators, fuse wire and plastic explosives we had used during a demolition course. On reflection, I should have returned these items, and I certainly had no criminal motive for keeping them, but, alas, I did. A twist of fate would lead me to meet an old friend a short time later, who wanted to protest at the way 20th Century Fox was desecrating a trout stream in the beautiful village of Castle Combe in order to shoot the movie, *Dr Dolittle*. I thought I had just the idea.

With my supply of explosives, I planned to use them to draw away the movie set security patrol, informing a supposedly friendly journalist of my plan to do so. Unsurprisingly, the journalist went straight to the police, who were soon lying in wait for the man the papers would come to call the 'Baronet Bomber'. I was very fortunate to escape jail, and although my fine of £500 stung at the time, I was also expelled from the SAS. My future father-in-law also rang my mother and threatened to call the police if I was ever to contact Ginny again. While it was an idle threat, and we would of course marry in time, that probably hurt me more than anything. It was certainly a very costly error and one that caused immense regret. Thankfully for the Ottomans, they were slightly better prepared than myself, and certainly did not phone ahead with their intentions.

As such, on 2 April 1453, over 200,000 Ottoman fighters, including the Janissaries and their firearms, surrounded Constantinople from the sea and around its landward walls. However, the Byzantines were expecting them. Having already destroyed the bridges across the moats, they had also closed the gates of the city, while a giant chain had been erected in the sea, which was designed to stop the Turkish fleet from getting too close. Upon the walls, and in the towers, were thousands of soldiers, armed with small cannons, javelins, arrows and even stones, to keep the invaders at bay.

Mehmed was nonetheless confident that he had the tools to smash through the walls, yet he hoped not to have to damage the city he aimed to make his capital. As such, he ordered Emperor Constantine to surrender immediately. If he did so, he would spare his citizens. Yet Constantine did not reply, believing his city to be impregnable. Mehmed now looked to prove him wrong.

With a cloud of black smoke, and a deafening roar, his cannons soon unleashed hell. Crashing giant cannonballs into the Theodosian Walls, they blasted away huge chunks, particularly at the Charisian Gate, which was severely damaged and collapsed a day later. The Byzantines desperately fired back with their own smaller cannons but soon had to stop as they found the vibrations damaged the fortifications. Just about managing to keep the Ottomans at arm's length, they spent the night trying to repair the walls. However, the Ottomans also dug underneath them, trying to cause them to collapse. In response, the Byzantines filled these underground tunnels with smoke, foul-smelling odours, or even water, to flush them out. Unable to go under, the Ottomans looked to go above, building a bridge made of barrels. Towering over the wall, it allowed many Ottoman soldiers to scale to the top before being shot down.

This relentless onslaught continued for the next six weeks. The Byzantine defence might have been weary, and short of men and materials, but somehow they repelled everything the Ottomans threw at them. And soon much-needed help arrived. Three Genoese ships sent by the Pope, as well as a vessel carrying vital grain sent by Alfonso of Aragon, managed to break through the Ottoman naval blockade and reach the city.

Infuriated by this breach, Mehmed now changed tack. He proceeded to build a railed road via which seventy of his ships, loaded onto carts pulled by oxen, could be launched into the waters of the Golden Horn. The Ottomans then built a pontoon, and fixed cannons to it, so that they could now attack any part of the city from the sea side, not just the land. This saw the Byzantines struggle to station men where they were needed, especially along the structurally weaker sea-side walls.

Time appeared to be running out for the city but, then, another reprieve came from an unexpected quarter. Back in Asia Minor, Mehmed faced several revolts. Knowing he had to return to quash any uprising, he offered Constantine a deal: if the emperor would pay an annual tribute of 100,000 gold bezants, the Ottomans would withdraw. But the emperor refused, being unable to raise such a tribute in any event, and still confident his forces, and walls, could hold out.

The Ottomans now faced a dilemma. At a war council, Halil Pasha, one of the Ottoman captains, urged Mehmed to forget the siege and return home:

> You have done your duty, you have given them a number of fierce battles, and every day great numbers of your men are killed. You see how strongly the city is defended, and how impossible it is to storm it; in fact, the more men you send to attack it, the more are left lying there, and those who manage to scale the wall are beaten back and killed. Your ancestors never got as far as this, or even expected to. It is to your great glory and honour that you have done so much, and this should satisfy you, without wishing to destroy the whole of your forces in this way.

But another captain, Zagan Pasha, appealed to the sultan to give it one last go:

> You have proved yourself the stronger. You have razed to the ground a great part of the city walls, and we shall break down the rest. Give us the chance of making one short sharp general assault and, if we fail, we shall afterwards do whatever you think best.

With this appeal, Mehmed vowed to launch one last assault to crush Constantinople once and for all, reminding his men of the riches the city contained and of the booty that could soon be theirs.

In the early hours of 29 May, Mehmed unleashed his final attack. His cannons once more blasted big chunks out of the walls. This was followed by a barrage of arrows from the cavalry and then wave after wave of infantry who, screaming their battle cries, attacked the walls. In the darkness, the church bells in the city rang out the warning that an attack was underway. Every man of fighting age, as well as women and children, raced from their beds and hurried to the walls to throw whatever they had at the Ottomans, as well as repair the crumbling walls.

Yet, after hours of fighting, Mehmed was dismayed to see that the walls had still not been breached. It was now time for him to turn to his elite troops: the Janissaries. If they failed, the conquest would be over.

Rather than rush towards the walls, as the previous troops had done, the Janissaries kept their ranks in perfect order, unbroken by the missiles of the enemy. Charging forward, wave after wave of these stoutly armoured men rushed up to the wall with their guns. Yet, as they tore at the barrels of earth that surmounted it, hacked at its supporting beams and placed ladders to scale it, shooting at the enemy above, they still could not make any headway. As the Byzantines held back the Ottomans' most elite troops, they now began to think that they could win. But fate was against them.

At the corner of the Blachernae Wall, just before it joined the double Theodosian Wall, there was, half-hidden by a tower, a small sallyport known as the Kerkoporta. Unbelievably, after weeks of bitter siege warfare, and centuries of armies being unable to breach the walls, the Janissaries found that someone had left the small Kerkoporta gate in the Land Walls open! Despite being the most formidable fortress in Christendom, which had seen off any number of invaders, it was an open gate that would prove Constantinople's downfall.

Pouring inside the city, the Janissaries fought and shot their way to the main gate. The Byzantines did all they could to stop them, knowing that should the gate be opened they would be overrun by tens of thousands of Ottomans. However, they were no match for the firepower of the Janissaries, who proceeded to shoot anyone in their way. Soon they reached the gates and allowed their comrades to flood into the decimated city.

With music playing and colours flying, the Ottomans stormed inside and killed anyone in their wake. Many of the city's inhabitants preferred to commit suicide rather than be subjected to the horrors of capture and slavery. Many others sought refuge in churches, and barricaded themselves inside. But these were obvious targets for treasures. After they were looted, the buildings and their priceless icons were smashed, while the cowering captives were butchered. Soon the emperor himself was dead, although legend suggests that he had been magically encased in marble and buried beneath the city, which he would, one day, return to rule again. Another story later circulated that two Turkish soldiers who claimed to have killed Constantine brought his head to the sultan, who had it stuffed and sent it to be exhibited around the leading courts of the Islamic world. Yet the emperor's fate was inconsequential. Constantinople was now in Ottoman hands.

In the afternoon, after allowing his men to loot and pillage, Mehmed finally entered the city himself, escorted by his finest Janissary guards. Calling an end to the pillaging, he declared that the Hagia Sophia church be immediately converted into a mosque. The city's role as a bastion of Christianity for twelve centuries was now over. Soon after, following the conquest of Mistra in 1460 and Trebizond in 1461, what was left of the old Byzantine Empire was absorbed into Ottoman territory.

Following such an historic outcome, the Ottomans manufactured gunpowder in Constantinople, as well as in the empire's provincial centres, shipping it all over their territory and allowing the Janissaries to continue to conquer new lands and empires. And soon they were casting their eyes at the Mamluk Empire.

While the Mamluks held the monopoly of the world's spice trade, and also ruled over Islam's holy lands, by 1516, they had failed to keep up with the change in technology. Predominantly relying on tactics and equipment perfected in the thirteenth century, their highly trained, horse-mounted archers at the core of their army were no match for the Janissaries' arquebuses. With their guns, the Janissaries made quick work of the Mamluks, killing the sultan and kidnapping the caliph, who was shipped back to Constantinople as a prisoner. As the Mamluk Empire fell, the Ottoman Turks held sway over almost all the Arab world. The conquest of Constantinople had transformed an essentially European power into a great Islamic-Mediterranean empire.

The advent of gunpowder had revolutionised European warfare, but it was those elite forces that not only adopted this new technology, but coupled it with ground-breaking tactics, who really reaped the rewards ...

12

THE LANDSKNECHTS

AD 1474

When the Burgundian Wars erupted in 1474, the Duke of Burgundy, known as Charles the Bold, had every right to believe his forces would prevail against the Old Swiss Confederacy and its allies. Fresh from its victory over the French, the Burgundian *compagnie* was considered one of the most feared, and most effective, ground forces in fifteenth-century Europe.

While its soldiers were instilled with rigid discipline, Charles had supplemented his forces with the best foreign mercenaries money could buy, including English bowmen, Italian heavy cavalry and German swordsmen. But, to the surprise of most, it was no match for the Swiss. In a series of upsets, Charles's forces were routed, most notably at the Battle of Nancy in 1477, where the Swiss destroyed Charles's armoured cavalry force and killed him in the process. The Swiss success seems to have been down to one factor – their proficiency with the pike, along with their infamous pike block formation.

While the use of pikes to fend off cavalry was common throughout the Middle Ages, such barricades were usually fixed in a defensive position. The Swiss pikemen changed this by introducing an offensive element to pike warfare. With the men in the front of the formation kneeling to allow the men in the centre or back to point their pikes over their heads, the well-drilled square could change direction very quickly, making it difficult to outmanoeuvre the block on horseback. It was also able to charge against the enemy, with levelled pikes and a co-ordinated battle cry. This was a game-changer for the military fields of Europe and such was their prowess that Swiss mercenaries, known as the Reisläufer, soon became sought-after soldiers for hire.

With Charles's death, his successor, Maximilian I, heir to the throne of the Holy Roman Empire, soon faced a challenge from France's King Louis XI and his successor Charles VIII, who each laid claim to the Burgundian legacy. In 1494, one year after Maximilian became ruler of the empire, Charles VIII commenced a series of attempts to invade Italy and take Naples and Milan. Aiding him in this venture was his very expensive army of mercenary Swiss pikemen. It was clear to Maximilian that, if he was going to protect his inheritance, he had to fight fire with fire.

The Burgundian Wars had shown that cavalry was virtually helpless against well-drilled pike formations. It was therefore pointless to expect the old, failed methods to prevail. Maximilian needed to somehow replicate the success of the infamous Swiss pikemen, and then beat them at their own game. To do so, he set up a mercenary army that sought to copy the Swiss in almost every way, which in time would be known as the 'Landsknechts'.

Fanatically studying how the Swiss recruited their troops, Maximilian contracted so-called 'gentlemen of war' to build a mercenary army on his behalf. Having accepted the appointment, and secured the means of finance, the colonel, or *Obrist*, would then send out his drummers to the lands of southern Germany, as well as the Helvetic Confederation, to beat for recruits.

One such *Obrist* was Georg von Frundsberg, lord of Mindelheim. Alongside his wealth and prestige, Frundsberg possessed a great personal charisma, and it was said that he was capable of raising armies of 20,000 men in a matter of weeks. It helped that the standard Landsknecht pay on offer was four guilders per month, which compared favourably with that of civilian occupations (a typical building worker in the city would earn two and a half guilders in 1515). Unlike others, Frundsberg also developed a reputation for ensuring that his troops were paid on a regular basis. All of this meant that he not only attracted the best recruits but also managed to keep his mercenary forces, and Maximilian, happy.

However, Frundsberg would not just accept any prospective recruit. A strict selection process based on physical fitness, equipment and social and economic status meant that only the very best men were recruited.

Having drummed in the villages for recruits, the men were instructed by Frundsberg to meet at a certain time and place for a muster parade. Here, they were ordered into two columns facing each other. At the end of the gap between them, an arch consisting of two halberds and a pike was erected. It was through this arch that each man had to pass a recruiting officer, who would check if they were of sound mind and body.

A figure that Frundsberg would have taken great care in selecting during this process was the ensign, who was regarded as only second in importance to the company captain. These men were usually chosen for their size, courage and skill in battle, as their role was to protect the company banner with their lives. Indeed, from the moment the banner was formally presented to an ensign during the muster, this flag, and its carrier, became the very symbol of the unit's 'manliness, courage and being' and the banner was never to fall into enemy hands. So important was the banner that, at the Battle of Marignano in 1515, a dead Landsknecht ensign was found with both his arms chopped off but with part of the banner pole still clenched between his teeth. If ever a man displayed true commitment to his job, this was it.

With Frundsberg having assembled his army, the men were immediately paid one month's wage before they assembled in a circle surrounding him. Frundsberg then informed them of their rights, duties and restrictions by reading out the 'Letter of Articles'. The articles consisted of a very detailed code of conduct and set out all the punishable offences, such as mutiny, unwarranted plunder, drunkenness on duty, having more than one woman following in the baggage train, and so forth. This was followed by an oath-taking ceremony in which every Landsknecht swore his allegiance to his cause, his emperor and his officers, and promised to abide by the laws set out in the articles.

Frundsberg now had to ensure that his men were well drilled and disciplined before they could be sent into battle against the French and their Swiss mercenaries. While the Swiss pike block had achieved tactical superiority during the Burgundian Wars, Maximilian was ready to go one step further. He realised that the introduction of the new arquebus, alongside the pike, could make for a formidable proposition. With this, he had soon devised a formation that he felt could make his Landsknechts more than a match for the Swiss.

Put simply, in a defensive position the pikemen and the halberdiers formed a solid square, and around this block stood a wall of arquebusiers affording protection from enemy pikemen. If the order to advance was given, a line of foot soldiers was strung out in front of the square, which was composed of either volunteers, prisoners who hoped to redeem themselves, or unfortunates who had been picked by lot. Known as the 'forlorn hope', it was their task to advance in front of the square with their pikes and two-handed swords, to stave off the oncoming enemy and hack their pikemen to pieces, so that their comrades would be able to penetrate the gaps they had made. The order would then be given for the regiment to form an *Igel* or 'hedgehog', either in a square or circle. The pikemen would now be at the front, levelling their weapons at an angle to take the oncoming cavalry, with the arquebusiers behind them opening fire.

Innovation such as this can soon catch up and even surpass the very best, as I have found to my cost. When I attempted to make a solo Antarctic crossing in 1996, I was known by the *Guinness Book of Records* to be the 'World's Greatest Living Explorer'. However, since my previous expedition, trekking in the snow had been revolutionised by one of my rivals, Borge Ouslund. Moving at a phenomenal speed, the secret to his success was the skilful use of a high-tech wind chute, which helped pull him, and his sledge, along. While I practised before I set out, I also had some real success using it in the Antarctic, once travelling over 117 miles in a day. By the twentieth day, I was already ten days ahead of my previous Antarctic journey, thanks to the addition of the kite, yet I was still struggling to keep pace with Ouslund. Alas, kidney stones were to cause me to abandon my attempt, while Ouslund continued on, making the crossing in a quite incredible fifty-five days, having sailed for over three-quarters of them. The Norwegian had adopted and perfected this revolutionary new feature in polar expeditions and had left his rivals in his wake.

This was certainly something Frundsberg aimed to do. Training his men day after day, his Landsknechts soon marched to war against the French, followed by an enormous train of wagons, supplies and hangers-on. Since each Landsknecht was a self-supporting entity, every soldier needed some personal support, which usually meant someone to cook for him, set up the camp, mend and wash his clothing, and nurse him when wounded or sick. While women and boys usually carried out this supporting role, some men were not so lucky. As such, a set of the articles of war from 1530 proposed that two or three women should also be engaged to be 'everyman's wife'. In other words, they were to be prostitutes available to the camp, who could charge two kreuzer for 'services of love'.

In the travelling camp, there was also a chaplain, a scribe, a doctor, a scout, a quartermaster, and a bodyguard of eight trustworthy men for Frundsberg. An independent group of officials was also responsible for maintaining discipline and ensuring that the Landsknechts abided by the articles. The most feared among them was the provost, who remained unimpeachable during his period of office. His retinue consisted of a jailer, a bailiff and an executioner called the *Friemann,* who was recognisable by his blood-red cloak and red feather in his beret, as well as his executioner's sword and hangman's rope, which hung from his belt.

As their wagons passed through towns and villages across the continent, the Landsknechts' dress caused quite a stir. Maximilian had made his Landsknechts exempt from the 'sumptuary laws', which dictated the colours and style of clothing each social class could wear, because, in his words, their lives were 'so short and brutish'. As a result, Landsknechts dressed in the most garish costumes. Slashed doublets, striped hose, tight or voluminous breeches and outrageous codpieces were all worn in a deliberate attempt to flaunt their status, intimidate their enemies, and shock civilians.

When their wagons finally reached the village of Wenzenbach, near Regensburg, they found the forces of the Palatine Ruprecht had taken up a defensive position on a hillock behind a wall of shields. It was time to put Frundsberg's Landsknechts to the test.

Firstly, Frundsberg sent out his 'forlorn hope' to engage the enemy. Ruprecht's cavalry made such quick work of them that, suitably emboldened, they charged through the ranks. Yet, as they did so, the Landsknechts immediately took up a defensive position, forming a wall of spikes, as well as guns, the likes of which had never before been seen on a battlefield. Before Ruprecht could react, his men were either impaled on spikes or shot off their horses. Over

1,600 men were killed in the ensuing slaughter. Maximilian's bold move to build such a mercenary army had been a roaring success. His Landsknechts developed such a reputation for unprincipled, ruthless violence that one chronicler remarked that the devil refused to let *them* into hell because he was so afraid of them.

However, the French knew well from their Swiss pikemen that all mercenaries had their price. As such, they managed to persuade some German Landsknechts to move away from Maximilian and fight on their behalf instead. According to Machiavelli, the French subsequently massacred Spanish forces at Ravenna in 1512, thanks to the stubborn resistance and fierce close-quarters fighting of these German Landsknechts. Maximilian was disgusted. Several days after Ravenna, he ordered all the German Landsknechts in the pay of the French to return home. All except 800 obeyed their emperor; with those 800 subsequently forming the nucleus of what became known as the 'Black Band'.

While Frundsberg continued to command his Landsknechts to more victories, with their pike block fast becoming impregnable, a new foe soon presented itself. In 1515, 20-year-old Francis I ascended the French throne. Around the same time, Charles V, Duke of Burgundy, the grandson of Maximilian and soon-to-be Holy Roman emperor, inherited the kingdom of Spain from Ferdinand. This suddenly posed a military threat not only to France but to the Pope as well, for the whole of southern Italy belonged to Spain and therefore most of Italy would soon be in Charles's hands. As such, when Maximilian died in 1519, Pope Leo X allied with Francis I, who had renewed his alliance with the Venetian Republic and Genoa. In addition, he had also recruited an army of 16,000 Swiss mercenaries. Just as his grandfather had done before him, Charles V looked to Frundsberg's Landsknechts to assert his authority.

Frundsberg and his Landsknechts soon joined the imperial forces in Italy, under the command of Prospero Colonna, and met the French at La Bicocca in April 1522. For the first time, Swiss and German mercenary pikemen would face each other in considerable numbers.

Colonna, the Italian commander in charge of the Spanish contingent, realised that the hunting lodge at La Bicocca presented a considerable defensive position, as a sunken lane ran between the bottom of a garden and the fields that separated the two armies. He subsequently ordered the bank to be built up into a rampart on the garden side. Then he positioned his arquebusiers on it in ranks four deep, along with several heavy cannons, with his German pikemen taking up the rear. As the Swiss advanced across the fields (with rocks and sand in their hands ready to throw at the enemy in order to blind them), their ranks were decimated by the murderous fire from the Spanish arquebusiers and artillery. And those who did succeed in reaching the lane suddenly found themselves in a deathtrap. The arquebusiers, who were so high up the bank that they could not be touched by the Swiss pikemen, proceeded to pepper them with gunshots. Frundsberg's pikemen then proceeded to rush down into the lane to finish them off. Some 5,000 Swiss, including twenty-two of their officers, were killed in the onslaught, including their commanders Albrecht von Stein and Arnold von Winkelried. The victory at La Bicocca had finally proved the superiority of the Landsknechts against the Swiss, just as Maximilian had dreamt.

By 1523, all France's previous successes in Italy had been undone. Matters were made worse for Francis when Francesco Sforza, an imperialist ally, now held the Duchy of Milan, long regarded as the rightful inheritance of France. After raising 40,000 men, Francis set off a year later to claim what he believed was his, soon taking Milan as well as a number of other highly fortified cities.

This turn of events was very serious for Charles, particularly after Pope Clement VII also turned his back on the empire, entering into an alliance with France and the Venetian Republic. Following this, a French army under the command of the Duke of Albany was allowed to pass through the Papal States unopposed. An army of German Landsknechts, under the command of Kaspar von Frundsberg, the son of the famous commander, and Graf Eitelfritz von Hohenzollern swiftly marched on Milan but they were soon forced to flee to the fortified town of Pavia, where they joined the old Spanish general Don Antonio de Leyva. Before long, the French began assaulting the southern walls and a siege was underway.

In January 1525, Charles desperately approached the ailing Frundsberg and begged him to once again raise an army to do battle with the French. Despite his ill health, Frundsberg agreed. He not only wanted to serve his master but also to help rescue his son at Pavia. Quickly, Frundsberg marched to Lodi, northeast of Pavia, where he joined Marx Sittich von Ems, who had also brought troops with him. Together their forces now numbered some 17,000 infantry and 1,000 horses.

Realising that the imperial relief force would soon be arriving, Francis moved his headquarters to the nearby park of Mirabello, securing a strong position between Pavia and the oncoming imperialist army. However, after three weeks of trench warfare, Frundsberg's men succeeded in making contact with the Spanish general Leyva in Pavia. Supplying him with much-needed ammunition and provisions, they also co-ordinated plans for the oncoming main assault.

Meanwhile, drenched by rain and decimated by sickness, the French army's morale was rapidly dwindling. On 20 February, 6,000 French troops even insisted on returning home, while 2,000 Germans deserted, thus reducing Francis's army to fewer than 20,000 men. At this, his generals advised Francis to withdraw, but he refused. The stage was set for a battle to the death.

At midnight on 24 February, the imperialist army, under cover of an artillery barrage and a noisy decoy, moved northward up the River Vernavola. Silently crossing the ford they reached the wall of the Mirabello park, which had thus far been protecting the French. Without drawing the attention of the enemy, the Spanish engineers worked through the night, forcing a 50-yard-wide breach in the wall.

At this, Frundsberg ordered seven companies of Landsknechts, numbering 2,800 men, to put white shirts over their armour (those who did not have white shirts were ordered to use paper) so that they could easily recognise each other in the darkness. Then, with three blasts from a cannon, Frundsberg gave the signal to those inside Pavia that it was time to move. While his men piled into the French outside the city, with their pikes and guns, Leyva and Kaspar led the charge from inside. Their shock attack swiftly succeeded in driving hundreds of the French into the River Tissino, where many drowned in their heavy armour.

With the French in disarray and trying desperately to stop the imperial forces from swarming through the breach in the park wall, Frundsberg's Landsknechts suddenly found themselves in battle with a familiar foe – the Black Band Landsknechts.

The subsequent fighting was ferocious. Packs of pikes stormed into the other, piercing skin and armour, while close-range gun-shots blasted opposing numbers to bits. But soon Frundsberg's men had the Black Band on the run. Upon seeing his best mer-cenaries in retreat, King Francis cried, 'My God! What is this?' Suddenly, his horse shot from beneath him and a mob of vicious Spaniards descended upon him. He was only saved by the speedy intervention of the imperial Spanish commander Charles de Lannoy, who granted him safe conduct from the field of battle.

In under two hours, 8,000 Frenchmen had fallen at the expense of only 700 imperialists. The defeat of the French at Pavia left Italy at the mercy of Charles V and proved beyond any doubt that the German Landsknechts were the best shock troops in Europe.

Francis was subsequently exiled to Spain and had to suffer the humiliation of complying with Charles's terms of surrender. These included pledging to renounce his claims on Burgundy, Italy and Flanders before he could return to his kingdom. This marked the lowest ebb in France's fortunes since the Battle of Agincourt, over a century earlier. It also signified the end of Swiss military supremacy.

However, no sooner was Francis reinstated at his court than he declared the terms of the surrender invalid and once again sought to take Italy. While Frundsberg's army of 12,000 Landsknechts prepared to march on Rome, news came that a peace treaty had been signed. But the Landsknechts were unimpressed. The treaty offered them a sum far less than they had been expecting for serving in a full campaign. With his troops threatening to mutiny, Frundsberg tried to placate them, but, exhausted by his extensive campaigning and now in ill health, he collapsed in front of his men and was to die soon afterwards. With the death of their beloved leader, discipline became virtually non-existent. Having been cheated of their full pay, the thought of rich plunder now drove the Landsknechts on towards Rome.

Within three hours, the whole of the Vatican had been taken. After the brutal execution of some 1,000 defenders of the papal capital and shrines, the pillage of Rome truly began. Churches and monasteries, as well as the palaces of prelates and cardinals, were looted and destroyed. Even pro-imperial cardinals had to pay to save their properties from the rampaging soldiers.

The so-called 'Sack of Rome' had major repercussions for Italian society and culture. While masterpiece sculptures and paintings were lost, the city's population dropped from some 55,000 before the attack to 10,000 afterwards, with an estimated 12,000 people murdered. Many imperial soldiers also died in the aftermath, largely from diseases caused by masses of unburied corpses in the streets. Pillaging finally ended in February 1528, eight months after the initial attack. When the city's food supply had run out, there was no one left to ransom. It would take Rome decades to rebuild, while it also did irreparable damage to the reputation of the Landsknechts.

Not only did the ultimatum 'No money – no Landsknechts' soon became one enshrined in infamy, but the body of pikemen and arquebusiers, who had been more than a match for the Swiss, was gradually superseded due to the progress of firearms. However, while the march of technology and military formations soon came to dominate the European battlefield, in Japan a group of elite warriors were achieving incredible results through nothing more than their bodies, and their ingenuity ...

13

THE NINJA

By 1562, Japan was in the grip of a civil war that had lasted for over 100 years. Where there had once been peace, warlords and violent militias now roamed the countryside in an orgy of destruction.

One such battle had ended in a year-long siege at Kaminogo Castle, an Imagawa clan outpost in the wilds of Japan led by Udono Nagamochi. While his rival, Tokugawa Ieyasu, had gathered outside his walls, along with a huge army, he didn't dare to attack.

Years earlier, Ieyasu had agreed to send his two daughters to Kaminogo in a peace deal with Nagamochi. This was standard procedure in a country where no one trusted anyone and was a way to help maintain allegiances. If everyone behaved, the families wouldn't be touched. But things had now changed. With Ieyasu and Nagamochi head to head once again, Ieyasu was well aware that just one wrong move would see his daughters murdered. It was checkmate. To break the impasse, Ieyasu would need to do something extraordinary.

However, while his samurai warriors excelled in battle, their skills were unsuitable for situations such as this. Their honour-driven code of *bushidō* (the 'way of the warrior') didn't allow them to use the tools of deception or dirty tricks. They were only able to face their enemy head-on in an honourable battle. In a situation that required unconventional guerrilla tactics, this wouldn't do. As such, Ieyasu began to look at another option: the shadow warriors, also known as the 'Ninja'.

The Ninja were the special forces of feudal Japan, who excelled in unconventional warfare. Renowned for deception, espionage and murder, they moved in small teams and exploited speed, aggression and surprise, all the while apparently being invisible. Such was their prowess that some regarded them as supernatural beings. But not all Ninja were created equal. It was the Ninja of Iga who were considered the finest purveyors of their craft.

Cut off from the rest of the country, Iga was situated on a high mountain plateau, where the river rolled through its valley, filling it with mist and giving it an otherworldly feel. Its people were also very different. Fierce, resourceful mountain folk, they were proud of their independence, and what they lacked in riches they made up for with their use of their surroundings, making a little go a long way.

It was in these rural hinterlands and highland villages that the art of the Ninja was perfected over many centuries. Most boys would spend their childhood learning the martial arts, as well as mastering the famous samurai sword, the spear, bow and, later in Ninja history, guns. They would also be expected to ride well and to swim. Yet perhaps their most vital lesson was learning ninjutsu, the art of invisibility, which encouraged them to blend into their surroundings before launching an attack. Any budding Ninja would also have to learn about such matters as explosives and the blending of poisons, and become an expert in fieldcraft and survival.

Sent out into the wilderness, the young boys were exposed to the freezing temperatures at high altitude and had to learn to live off the land for an indefinite period. In this environment, the Iga Ninja became an expert at optimising their diet to resist cold and hunger. Hunting was also a vitally important training technique. This taught the Ninja how to track and study the habits, movements and routines of his prey, all without making a sound, before striking when it was at its most vulnerable.

While the Ninja of Iga were considered the best in Japan, among them was one member who towered above all others – a 20-year-old rising star called Hattori Hanzo. A former samurai, Hanzo had mastered the Ninja craft like no other. Said to be fearless during operations, some also added teleportation, psychokinesis and precognition to his ever-growing list of skills. It was for this reason he was nicknamed the 'Demon'. Yet there was one particular skill in which Hanzo was said to excel – penetrating castles and freeing hostages. Knowing this, Ieyasu hired Hanzo and his men to rescue his daughters from Kaminogo Castle.

First, Hanzo needed to gather intelligence in order to decide his strategy. Like the Assassins before him, as soon as he arrived in the surrounding area he planted his Ninja into the local villages. There, they posed as a sect of Zen monks, who played the flute and provided prayers and blessings. This ruse had proven successful in the past and Nagamochi also fell for it, inviting Hanzo's sleeper agents into his castle undetected.

With the intelligence they provided, combined with his unrelenting surveillance of the castle, Hanzo soon established the nature of its defences and its routines. He decided to enact a pincer movement, using his forces on the inside and outside to rescue Ieyasu's daughters, and take the castle, all in one swoop.

Now Hanzo needed to wait for the right conditions in which to strike. The Ninja never moved on a full moon in order to avoid detection. Tradition dictated that they had to wait for eight days after the moon, or eight days before it, to ensure maximum darkness. Like most Ninja, he also followed Shintoism, a religious practice focused on the natural world. For the Ninja going on operations, it was crucial to become at one with nature and blend into their environment. While they would utilise camouflage and clothing, much like the Navy SEALs and SAS do today, they also eliminated any shapes that could give them away, removed shine from their forehead, nose or weapons, and ensured any shadow or silhouette was kept to a minimum. Other tricks, such as leaving no sign of disturbance and walking lightly, almost like a ballet dancer, to avoid making any noise, were also crucial.

Hanzo also used an ancient practice called Kuji Kiri, one of the spiritual arts that honed concentration and produced complete psychological detachment from fear. A Ninja would form nine different shapes with his hands in order to make his mind strong, harnessing power, energy, harmony, healing, intuition, awareness, force, creation and Zen as he did so.

With everything in place, Hanzo stood outside the castle walls in the darkness. Dressed all in black, with only his eyes visible, he forced some iron spikes into the clay, then steadily began to climb to the top. But scaling walls in silence and secrecy was the easy part. Hanzo now had to move unseen through a castle full of soldiers on patrol. Thankfully, his sleeper agents had already fed him details of all the patrol movements.

Gliding through the shadows, running on the balls of his feet to eliminate sound while crouching low, he weaved in and out of columns, balancing his sword's scabbard on the tip of his blade as he did so. This extended his range of feel by a good 6ft. If it should touch an enemy in the darkness, Hanzo would let the scabbard fall before plunging his sword into them.

Following one sentry from behind, careful to ensure the sound of his footsteps matched his enemy's, he suddenly whipped the base of his fist in an upwards stroke to the base of the sentry's neck. The blow was so powerful it lifted the sentry's skull free of his cervical vertebrae and severed his spinal cord. Death was instantaneous. When any guard should move his way, he hid in plain view, in the shadows, not moving, not daring to stare at the enemy in case they might sense his presence. But there was one fight he couldn't avoid. Outside Ieyasu's daughters' room were two guards. To get inside, Hanzo needed to take them down quickly, and quietly.

In a flash of movement, he grabbed one from behind and slit his throat with a dagger, then, as the other turned, he slashed at him with his shorter ninjato sword, also known as the 'sword of darkness'. At this, he signalled to his sleeper agents that it was time to make their move. They proceeded to set the castle alight, as a signal to Ieyasu's forces that they could now attack.

As hordes of samurai overwhelmed the castle defences, Hanzo looked through the crack in the door to ensure there were no more guards in the room. Not taking any chances he stood to the side and slowly opened the door, being careful to avoid his shadow moving into the room, giving him away. But the room was clear, with only Ieyasu's two daughters present. As mayhem erupted inside and outside the castle, Hanzo proceeded to smuggle them to safety, while Ieyasu defeated his rival. This operation not only cemented Hanzo's reputation as the best Ninja in Japan, it also saw him build an alliance with Ieyasu that, in years to come, would prove to be crucial.

Seventeen years after the siege of Kaminogo Castle, Japan continued to be torn apart by feuding warlords. However, one aimed to bring the country under his sole command, by any means possible. The genocidal maniac, Oda Nobunaga, and his gigantic army crushed all who stood in their way, gaining a reputation unmatched in all of Japan's bloody history. Despite this threat, the people of Iga refused to be cowed.

Faced with such disobedience, Nobunaga ordered his forces, led by his son, Oda Nobukatsu, to invade Iga. Nobukatsu seemed to think that all his ruthless band of samurai had to do was enter Iga and all would bow before them. He could not have been more mistaken. Walking through the main forest path of Iga, Nobukatsu had underestimated his foe, overestimated his own capabilities, while he did not know the lie of the land. Each mistake would prove to be fatal, as watching the samurai's every move from the shadows was Hanzo and his Ninja.

Before making their move, the Ninja waited for the valley to become enshrouded in mist from the river. Thus, in the early morning gloom, Hanzo called for them to strike. Seemingly appearing from out of nowhere they attacked from all sides, overwhelming the confused samurai, causing chaos. Such was the panic and confusion that the samurai began to blindly lash out with their swords, cutting each other down in the process. Soon the only survivor was Nobukatsu, and rather than face the Ninja, or fall on his sword, he made a run for it. The almost total annihilation of Nobunaga's samurai was one of the Ninja's greatest triumphs, but it would also unleash forces that would test the men of Iga as never before.

Within days, the Ninja spy network reported that Nobunaga was mustering the biggest army Japan had ever seen. Columns were now streaming from every region with one order: destroy Iga. The Ninja knew they couldn't defeat so huge an army. Again, taking their cue from the Assassins, there was only one solution – Nobunaga had to be assassinated.

The man appointed for this high-risk job was Ishikawa Goemon. To the people of Iga, Goemon was almost a Robin Hood figure. Renowned for stealing from the rich to give to the poor, he was also regarded as an exceptional Ninja. However, his most important qualification for this operation was that it was deeply personal for him. In 1573, his whole family had been murdered by Nobunaga's men and since that fateful day he had sworn revenge.

But killing Nobunaga was the toughest assignment ever faced by a Ninja assassin. The warlord was constantly surrounded by bodyguards, while he also employed Ninja from Iga's rivals. What's more, Nobunaga knew that Iga would be plotting to kill him, so he was ready for them.

Again, this was a situation that could not be met head-on. It would need cunning and thinking outside the box. As such, Goemon believed that poisoning Nobunaga might be his best option, especially if he wanted to avoid detection and protect the people of Iga from retribution. In such situations, the Ninja would typically poison their target's food or water supply. But Nobunaga employed a host of food tasters to protect him from this very threat. Yet Goemon came up with a simple, but ingenious, plan.

Just as Hanzo had proved at Kaminogo Castle, infiltrating a fortress was easy work for a well-prepared Ninja. Choosing his moment carefully, Goemon ensured he blended in with his surroundings, and using the intelligence provided by a sleeper cell he also knew what to avoid and where to go.

Navigating his way across the roof of the Azuchi Castle fortress, keeping low to avoid any silhouette, he entered above the royal apartment complex. Making a small hole in the ceiling, above where Nobunaga was sleeping, Goemon took out a long piece of string, gently threaded it through the hole, and left it hovering just above his target's open mouth. Taking out his bottle of poison, he applied a few drops to the top of the rope and watched as the deadly liquid slowly trickled downwards. But, at the very last moment, just as it was about to drop into Nobunaga's open mouth, he turned his head, leaving the poison to splash against his cheek. Waking with a start, Nobunaga saw the piece of string, as well as Goemon above him.

With the alarm raised, Goemon knew his mission had failed. All he could do now was try to escape the palace alive using all of his Ninja skills. As guards gave chase, he threw small steel spikes, known as caltrops, on the floor. This slowed some of his pursuers but the samurai bodyguard still gave chase. Reaching into his pockets, Goemon grabbed a handful of shuriken Ninja stars and launched them at his enemy. The pointed ends embedded themselves into his pursuers' flesh and sent them writhing in agony to the ground.

Yet Goemon was still not safe. As he turned a corner, he appeared to have entered a dead end. Hearing the ever-nearing footsteps of the samurai, and knowing he couldn't fight them all, he took out a special mix of gunpowder, threw it to the floor and lit it. Soon the area was a sea of smoke, and Goemon was able to creep right past the samurai undetected, and finally make good his escape.

My own family was once involved in a high-profile assassination. In 1327, my ancestor Roger Mortimer was accused of murdering Edward II by sticking a red-hot poker into His Majesty's anus. Mortimer was apparently the lover of the king's wife, Queen Isabella, and this was a chance for them to rule together. However, Roger was hanged, drawn and quartered for his crimes, after being arrested by Edward III in 1330.

Unlike Edward II, Nobunaga survived his assassination attempt and now went on the warpath. Taking personal command of his massive army, and enlisting rival Ninja from far and wide, he unleashed them against Iga, giving one order: spare no one.

Soon Iga was overwhelmed by over 60,000 of Nobunaga's men, who destroyed all homes and villages, while they also slaughtered Ninja, women, children and monks. This was nothing less than an attempted genocide, the likes of which Japan had never seen. When he was finished, the Iga people had been almost totally eradicated.

Thankfully, some Ninja, including Hanzo, escaped to the hills. However, hunted down by Nobunaga's thugs and with no land to return to, they were unable to offer their services elsewhere as everyone was aware that the Iga Ninja were wanted men. All they could seemingly hope for was to survive day by day.

Just as all seemed lost, Hanzo heard some incredible news.

In a palace coup, Nobunaga had been assassinated by his vassal Akechi Mitsuhide. His successor, Toyotomi Hideyoshi, now sought to wipe out all of Nobunaga's allies, one of whom was Tokugawa Ieyasu.

While many of the Ninja thought they could now relax and maybe even return home, Hanzo spotted an opportunity that could see them rise again like never before. With the once powerful Ieyasu hunted down and unable to get home, Hanzo told his fellow Ninja that they must help him. While Ieyasu was an ally of Nobunaga, he was also a friend of the Iga Ninja. If they could help him now, during his moment of greatest need, then the Ninja would have a powerful friend for life. His men reluctantly agreed. This could be their salvation.

Making contact with Ieyasu, Hanzo and the Iga Ninja helped him to escape back home to Mikawa, through the perilous mountain regions of Kōga and Iga. But lying in wait for them was a hunter-killer unit of samurai, who aimed to slay Ieyasu and finish off the Ninja once and for all. In the ambush in the snow, a vicious fight broke out. While the samurai hacked and swiped at them with their swords, the Ninja were far too quick and agile. Dodging the blows, they proceeded to take down the samurai one by one. Forcing the samurai to retreat, the Ninja were able to guard Ieyasu all the way back to Mikawa.

In the years that followed, Ieyasu regrouped and rebuilt his power base with the help of Hanzo and his Ninja, who now guarded him at all times. Incredibly, in 1600, Ieyasu assumed complete control of a united and peaceful Japan, ushering in a golden shogun dynasty that would last for the next 265 years. It also guaranteed the survival of the Iga Ninja, who continued to guard Ieyasu and his descendants over the coming centuries.

To this day, the legend of Hattori Hanzo and his Iga Ninja lives on. In Tokyo's Imperial Palace (formerly the shogun's palace), there is a gate called Hanzo's Gate; there is also the Hanzomon subway line, which runs from Hanzomon Station, in central Tokyo, to the southwestern suburbs.

You can also visit Hanzo's remains, which rest in the Sainen-ji temple cemetery in Yotsuya, Tokyo, where his favourite spear and ceremonial battle helmet are also on display. The spear, originally 14ft long, was given to him by Ieyasu and, though it was damaged during the bombing of Tokyo in 1945, it remains a major tourist attraction. In popular culture, Hanzo and the Ninja continue to endure, being the subject of countless books, movies and computer games throughout the world, ensuring that their legacy will never be forgotten, while also continuing to serve as inspiration to today's special forces.

But, back in England, a civil war had also broken out. To settle it would involve one of the country's most remarkable men building a new army, the likes of which had never been seen before . . .

14

CROMWELL'S NEW
MODEL ARMY

AD 1644

By 26 October 1644, the English Civil War had raged for two years. On one side of the divide were King Charles I and his Cavaliers. On the other were the Roundheads of Parliament. But, with Parliament's 19,000-strong army gathering around the Royalist outpost of Donnington Castle, their superior numbers looked set to put an end to matters once and for all. However, the king had defied Parliament right from the start of his reign. And now, despite being surrounded, he looked to do so once again.

When Charles ascended the throne in 1625, he had immediately butted heads with Parliament when it refused to raise taxes in order to fund wars in Spain and France. While Parliament did not believe such conflicts were worthwhile, the English armed forces were also in dire need of investment, with one general stating, 'The number of lame, impotent and unable men, unfit for actual service, is very great.'

Moreover, England had been at peace for so long that men who had previously fought in wars were now too old to be of any use. On the other hand, the young had no knowledge of war and were therefore grossly unprepared. The training of the current armed forces was also pitiful. While men who held estates of certain value were bound to provide trained 'bands', in different regions of the country, they only met to drill once a month during the summer. In most cases, these get-togethers seemed to focus more on drinking in local taverns than any actual military training. As a Colonel Ward later commented, 'As trainings are now used, we shall, I am sure, never be able to make one good soldier.' It was said that London was the only region with trained bands of any consequence, as they hired expert soldiers to instruct them, but they still fell well short of the best European armies of the age.

Of further concern was the decrease in horsemen and the decline in the breeding of horses. A famous soldier of the time, Sir Edward Harwood, said of this: 'And for horse, this kingdom is so deficient that it is a question whether or not the whole kingdom could make 2,000 good horse that might equal 2,000 French.'

For these reasons, Parliament refused to give way to Charles's demands to wage war. In response, Charles dissolved Parliament a number of times in an attempt to get his way. Two disastrous wars with the Spanish and French followed, resulting in Charles again dissolving Parliament in 1629 in an attempt to centralise power.

However, eleven years later he would have to go crawling back. In the interim, religious disputes between Protestants and Catholics had continued to tear the kingdoms apart, and, when war erupted with Scotland, Charles had no choice but to recall Parliament in 1640 in order to fund the ongoing conflict. To his surprise, Parliament again refused to offer any support. Once more, his underfunded, undisciplined, inexperienced forces were humiliatingly defeated on the battlefield. Such was the sorry state of affairs that a band of Devonshire men had even murdered their lieutenant because they suspected him of being a Papist. In short, the English army was more an armed mob than a professional outfit.

Matters worsened for Charles in 1641 when Catholic forces revolted in Ireland. Once again he went to Parliament to ask for taxes to be raised and still he was refused. Relations were now so bad between them that Parliament had become concerned that, if they were to give Charles the funds he required to build an army, he might then use it against them. With this in mind, Parliament instead raised its own army to fight the Irish Catholics. For Charles, this was nothing more than an affront to his sovereignty.

Soon Parliament and the king were at loggerheads, which saw MP John Pym take the highly provocative measure of placing guards at Westminster. As plots of rebellion grew, the king marched to Parliament with a body of soldiers, intending to arrest five leading members of Parliamentary opposition. Yet the five MPs managed to escape, leading to the king's authority being further damaged.

Sensing that civil war was now inevitable, the king fled to York, while his Catholic wife, Henrietta Maria, sailed to the Netherlands in order to buy arms and armour. Thus began the English Civil War, with both sides now having to improvise an army out of masses of untrained men.

When the king, through his Commission of Array, and Parliament, through its Militia Ordinance, summoned the trained bands to fight on their behalf many did not know which side to choose. Indeed, many had no interest in the war. One such family that was split on the matter was my very own. William Fiennes, Lord Saye, and his three younger sons sided with Parliament. Fiennes was even made Lord Lieutenant of Oxfordshire, with the power to call out and command militia, while his three boys went on to become Roundhead generals. All were subsequently present at the first real battle of the civil war in Edgehill, upon which they lost the family's castle at Banbury days later in a crushing defeat. Yet, while they put their lives and fortunes on the line for Parliament, Fiennes' eldest son, James, tended to side with the

Cavaliers. Moreover, one of William Fiennes' daughters even married into a Royalist family. One can only imagine the awkward conversations around the Fiennes dinner table at this point.

However, even when a trained band did choose a side, many often refused to march out of their county. Each regional association pictured itself in the mould of a self-contained unit rather than as part of a wider cause. When Sir William Waller discovered that his London-based units were refusing to campaign further afield, he wrote, 'An army compounded of these men will never go through with your service, and till you have an army merely your own that you may command, it is in a manner impossible to do anything of importance.'

On the rare occasion they could be persuaded to fight beyond their counties, they were almost impossible to command, which moved Waller to say: 'They are so mutinous and uncommendable, that there is no hope of their stay ... Such men are only fit for the gallows and a hell hereafter.'

It was only the trained bands of London that could be relied upon and as such Parliament turned to them in every emergency. But, while discipline was always an issue, most in the bands seemed too accustomed to good food, and good beds, to endure too much hardship on any long campaign.

As the trained bands in general were unserviceable, each party attempted to raise an army by voluntary enlistment, urgently appealing for men and horses. Yet the zeal of their supposed supporters was insufficient to fill the ranks and both sides eventually had to resort to impressment.

With the men's lack of discipline, little faith in their cause and haphazard pay, two years of indecisive fighting followed, and the conclusion of the war seemed no nearer. It appeared that each side could raise a body of troops strong enough to defeat its opponents in battle, but neither side could then keep them together to carry a campaign to a triumphant and decisive conclusion.

Yet, after routing the Royalists at the Battle of Marston Moor in 1644, Parliament's vastly superior forces of 19,000 men, compared to the Royalists' 10,000, now looked to finish things at Newbury. While Edward Montagu, 2nd Earl of Manchester, was in supreme command of Parliament's forces, it was the commander of the cavalry whom the Royalists feared more than any other. His name was Oliver Cromwell, and, despite having no prior military experience before the war began, he had gained a reputation as a formidable commander, whose men were among the best trained and most disciplined in all of England.

Unlike many of his peers in the House of Commons, the middle-aged MP for Cambridge did not come from noble stock. Rather than relying on wealth or privilege, it was through sheer force of personality and fierce intelligence that saw Cromwell rise to serve in Parliament.

As was the case with many MPs, on the outbreak of war, Cromwell was charged with raising an army from his native East Anglia. Subsequently assembling a cavalry, Cromwell was more interested in the ability of his men, and their belief in the cause, than their social origins, stating: 'I'd rather have a russet-coated captain that knows what he fights for than what you call a gentleman and nothing else.'

However, his religion also played a big role when selecting his men. Before serving as an MP, Cromwell had appeared to be washed up. Suffering from depression, and in poor financial circumstances, he subsequently underwent a religious conversion that was to propel him from middle-aged obscurity to national power. Inspired by his fierce Puritan beliefs, Cromwell ensured that his ranks were full of like-minded men. The Presbyterian minister Richard Baxter has said of this:

At his first entrance into the wars, being but a captain of horse, he had a special care to get religious men into his troop: These men were of greater understanding than common soldiers, and therefore were more apprehensive of the importance and consequence of the war; and making not money, but that which they took for the publick felicity, to be their end, they were the more engaged to be valiant; for he that maketh money his end, doth esteem his life above his pay, and therefore is like enough to save it by flight when danger comes ...

While the Royalists might show devotion to their king, Cromwell believed his men would show far more loyalty, and determination, to their God. Yet such was his force of personality that many were more than happy to be led by Cromwell himself.

Just a few months previously, Cromwell's cavalry had displayed its brilliance at the Battle of Marston Moor. While Royalist cavalry tended to pursue fleeing enemies off the battlefield, Cromwell's cavalry did the opposite. Instead, they broke the enemy formation, quickly reformed, and were ready for their next target. Such tactics saw Cromwell destroy Prince Rupert's infamous horse, and saw his men, now named the 'Ironsides', lauded throughout the land.

With far superior forces at Newbury, including Cromwell's Ironsides, it looked to be an opportunity to deal the Royalists a decisive blow. However, what followed was a disaster. With ill-disciplined bands, and a chaotic command structure, Parliament's attempt to outflank the Royalists was a disjointed mess. Even Cromwell's cavalry failed to live up to its reputation as it was uncharacteristically drawn into battle on boggy ground, where it quickly floundered. Matters weren't helped either by the Earl of Manchester delaying his attack until it was too late, and then allowing King Charles to slip past them at night. Cromwell, frantic at this lost chance, set out in pursuit, but Manchester refused to support him with his infantry, claiming they were exhausted. Against all odds King Charles made it back to Oxford and was once again able to regroup, with the war set to rumble on.

Cromwell was furious at the Parliamentarians for failing to press home their significant advantage. He put the blame on a disorganised army and cowardly, unprofessional commanders, not least the Earl of Manchester, who had let the king and his army escape. This issue soon exploded. Never one to hold his tongue, Cromwell made serious accusations against Manchester, charging him with being less than enthusiastic in his conduct of the war, as he was a Presbyterian and therefore inclined to favour peace with King Charles.

With Cromwell and Manchester at each other's throats, the Parliamentarian Eastern Association of counties announced that they could no longer meet the cost of maintaining their forces. This represented a disaster for Parliament since this band provided around half of its field force. Things could clearly not continue in this manner. After years of underinvestment, a full-scale review of Parliament's forces was now urgently required.

Unsurprisingly, when Parliament convened on 9 December 1644, the ever-combative Cromwell had plenty to say on the subject, telling the Commons:

It is now a time to speak, or for ever hold the tongue. The important occasion now is no less than to save a nation out of a bleeding, nay almost dying condition, which the long continuance of this war hath already brought it into; so that without a more speedy, vigorous and effectual prosecution of the war – casting off all lingering proceedings like [those of] soldiers-of-fortune beyond sea, to spin out a war – we shall make the kingdom weary of us and hate the name of a Parliament.

He then proceeded to list all the issues, as he saw them, with the current Parliamentarian forces:

Enlargement upon the vices and corruptions which were gotten into the army, the profaneness and impiety and absence of all religion, the drinking and gaming, and all manner of licence and laziness; and said plainly that, till the whole army were now modelled and governed under a stricter discipline, they must not expect any notable success in anything they went about.

In his mind, it was clearly time for a reorganisation of the army and a change in its commanders. The MP Zouch Tate then put forward a controversial suggestion that would have far-reaching effects. He proposed that, 'during the time of the war no member of either House shall have or execute any office or command, military or civil, granted or conferred by both or either of the Houses [Commons or Lords]'.

This was to be known as the 'Self-denying Ordinance'. It prohibited anyone who was serving as an MP from also serving as a general in the army. This was highly controversial but soon the House of Commons, and the less receptive House of Lords, realised it had no choice but to totally reform its armed forces and embrace what was to be known as the 'New Model Army'. My ancestor, William Fiennes, Lord Saye, was said to be particularly crucial in persuading the Lords to back the ordinance.

With the Self-denying Ordinance coming into effect, Members of Parliament reluctantly renounced their individual military commands, including the Earls of Manchester and Essex, as well as Oliver Cromwell himself. As for the army, several major changes were now implemented.

Raising funds was crucial if Parliament was to build a professional army. Therefore, a levy of £6,000 a month was issued on all the districts under the control of Parliament. This allowed all soldiers to be paid reasonably well by the standards of the time, and, perhaps of more importance, they were finally to be paid promptly. This helped to reinforce the sense of discipline vital to the army and to maintaining morale. Men knew that their pay was coming and this ensured they remained motivated and focused.

Central reorganisation also created the opportunity to resupply the army, providing troops with up-to-date equipment. For the first time, infantry regiments were uniformly equipped and all given adequate weapons and armour for the job. Alongside this, a single uniform of red jackets was introduced – the first time in English history that an army had shared one uniform. It was a step that also helped create an air of professionalism, although red was chosen because it was the cloth that could be obtained cheaply in large quantities.

In a further bid to ensure a unified and motivated front, the officer corps were mostly inclined towards Puritanism and adopted a rigid stance against the king. As with Cromwell's Ironsides, this allowed them to present a united front and push the enemy hard.

Where armies had previously answered to a haphazard network of commanders, all now answered to one command structure, which itself was ultimately answerable to Parliament. And the man Parliament chose to lead its New Model Army was Sir Thomas Fairfax.

Commander of the Northern Association, Fairfax had not only fought an effective war for Parliament in the north, but he was also regarded as a competent officer of unquestionable loyalty. Indeed, as the son of Lord Fairfax, Sir Thomas was also a good political choice. The nobles in the House of Lords had grown increasingly concerned at their waning power over the military, but Sir Thomas was seen to be one of them. He was close enough to the peerage via his father, but by virtue of not being a member of the House of Commons he was also still eligible to lead the army. This satisfied both Houses.

As for the army, Fairfax was certainly someone the men under his command could respect. A fully trained soldier, who had served in the Dutch army, he was also a blunt-speaking Yorkshireman with no airs or graces. His men knew that he would stand shoulder to shoulder with them through the hottest assault, while they also trusted his ability to lead.

As Cromwell had urged, positions in the New Model Army were no longer to be filled by the power of privilege, social standing or wealth. Instead, they were to be filled by those who were best for the job. So, while Fairfax could pick any colonels and other regimental officers he desired, he still had to submit his proposals to both Houses for their approval. In any event, any officers felt to be inferior were now discharged. Amazingly, some corporals and sergeants, unable to find employment elsewhere, stayed on to serve as ordinary soldiers.

With the Houses approving his list of commanders, Fairfax now turned his attention to putting together his army. Yet, after recalling all the armies previously commanded by the likes of Essex, Manchester and Waller, the numbers were insufficient. An ordinance was subsequently introduced to impress 4,000 more men. This soon saw the New Model Army consist of 22,000 soldiers, comprising eleven regiments of cavalry, each of 600 men; twelve regiments of infantry, each of 1,200 men; and one regiment of 1,000 dragoons.

By the summer of 1645, with the New Model Army assembled, Fairfax saw his chance to put his troops to the test. King Charles had split his forces to try to recapture the north of England while also retaining the south. This meant the king's defences around his base in Oxford would be depleted. For Fairfax, this was an opportunity too good to turn down.

The Parliamentarians besieged Oxford and the king's army raced to its assistance in large numbers. Suddenly, Fairfax realised that he would be outnumbered if he did not act quickly. He required the help of Oliver Cromwell, but under the terms of the Self-denying Ordinance the MP was not allowed to serve, even though he had continued to fight in the interim. Despite this, Fairfax urgently wrote to Parliament and urged it to appoint Cromwell as temporary Lieutenant General of the Horse, stating:

> The general esteem and affection which he hath both with the officers and soldiers of this whole army, his own personal worth and ability for the employment, his great care, diligence, courage and faithfulness in the services you have already employed him in, with the constant presence and blessing of God that have accompanied him, make us look upon it as the duty we owe to you and the public, to make it our suit.

In the circumstances, Parliament agreed to make an exception. Cromwell was to serve under a series of three-month commissions, which could be extended or terminated if the House so wished.

At six o'clock on 13 June, as Fairfax called a council of war at his headquarters in Kislingbury, cries suddenly erupted: 'Ironsides ... Ironsides has come.' Cromwell and 600 of his men had triumphantly arrived, having ridden with utmost haste out of the association counties. With Cromwell's arrival, the battle plans now moved on apace. Drums were ordered to beat to assemble the infantry while trumpets called out for troopers to horse, with the whole army ready to move on the small market village of Naseby.

Meanwhile, King Charles had called a council of war of his own. There, a debate erupted with Prince Rupert about whether to engage the New Model Army or retreat to fight another day. The decision was eventually made to go into battle, particularly as the king reasoned that the New Model Army was raw and untested in the field against an army of the king's experience. This was a once only opportunity to destroy it before it had a chance to grow.

Around five the following morning, Parliament's forces reached Naseby, which was covered in a thick mist, making visibility poor.

Despite this, across the field they could make out the Royalist army waiting for them on top of Dust Hill. Cromwell checked out the lie of the land for attack. He realised that his 3,500 cavalry, in their coats of buff leather, distinctive pot helmets, pistols and cut-lass swords, were facing a patch of extremely wet and boggy land. Remembering Newbury, he recognised that this would be deadly to any charge. He also didn't want to lead his men up Dust Hill to fight, as this would give the Royalists a significant advantage. Turning to Fairfax, he pointed to a piece of ground called Red Hill and said: 'Let us, I beseech you, draw back to yonder hill, which will encourage the enemy to charge us, which they cannot do in that place without absolute ruin.'

The consequent leftward movement of the New Model cavalry caught the eye of Prince Rupert. Fearful of being outflanked by superior numbers, and eager to attack, he suddenly unleashed the battle cry, 'Queen Mary!', an anglicised tribute to Henrietta Maria. In response, the New Model Army roared, 'God and our strength!' At this, both sides flew into battle.

After a volley of musket fire, the two armies were quickly upon one another, fighting hand-to-hand. Such was the ferocity of the Royalist attack that the vastly superior Parliamentarian infantry was at first pushed back. The king's secretary, Sir Edward Walker, described the scene:

> The foot on either side hardly saw each other until they were within carbine shot, and made only one volley, our falling in with sword and butt-end of musket did notable execution, so much as I saw their colours fall and their foot in great disorder.

Prince Rupert's cavalry had the upper hand but they then made a fatal mistake. Having routed a section of Parliamentarians, they chased it off the battlefield until they reached the Parliamentarian baggage train some 15 miles away. In comparison, Cromwell and his men, including my ancestors Colonel John Twistleton and Colonel John Fiennes, stayed on the field and regrouped. Such actions signified the newfound professionalism of the New Model Army and would be the key to its success.

Now outnumbering the remaining Royalist cavalry by more than three to one, Cromwell invited their disorganised horse to charge uphill. The Royalists took the bait. As they approached, Cromwell's infamous Ironsides charged through musket smoke and opened fire, before cutting the Royalists down with vicious swipes of their swords. After a brief but bloody encounter, the Royalist cavalry was trounced.

Cromwell now turned his Ironsides to outflank the Royalist infantry. Attacking them from the rear, they caused the Royalist troops to panic. Some tried to run, while others surrendered. Most were ruthlessly killed, with no mercy shown.

When the king saw the battle being lost, and his throne along with it, he rode forward with the intention of leading his own lifeguards into the fray. The Earl of Carnwath suddenly grabbed his horse's bridle and screamed: 'Would you go upon your death!' Without the king, there was no cause left to fight for. Facing the inevitable, Charles had no option but to flee to safety.

As the smoke cleared from the battlefield, the picture was one of complete ruin for King Charles. More than 1,000 of his men had been killed, with many more left mortally wounded on the field, while 5,500 had been taken prisoner. The cream of the king's army, and most of his veteran soldiers, had been lost.

In one day, the New Model Army had reduced the king's fortunes by a far greater degree than the old armies had achieved over the three preceding years. Unlike previous battles, no soldiers of the king had been allowed to withdraw unmolested from the field. There had been none of the almost farcical backing off to meet at another time, another place. Moreover, instead of chasing the enemy off the field, the men had remained and continued to fight.

In May 1646, while King Charles's decimated forces fought on forlornly, the king had no choice but to surrender himself to the Scottish army, in order to remain out of Parliament's hands. However, the Scots soon sold him back to the English and he was subsequently put on trial. While he awaited the verdict, William Fiennes, despite having supported Parliament during the war, was the only peer who paid for the royal children to be well looked after while their father was away. Charles was found guilty of treason, and executed on 30 January 1649. For the first, and only, time, England was a republic. Meanwhile, Cromwell and his New Model Army achieved yet another famous victory at the Battle of Worcester in 1651, where three Fiennes were present, helping to force Charles II into exile.

As well as earning acclaim as a military commander, Cromwell continued to be an outstanding politician, being sworn in as Lord Protector on 16 December 1653. Unlike the king, the Lord Protector was to seek the majority vote of a Council of State and supposedly abide by it, although it was to all intents and purposes a military dictatorship. Cromwell's power in such a position was also buttressed by his continuing popularity among his New Model Army.

While Cromwell had two key objectives as Lord Protector – to heal the nation after civil war and to ensure spiritual and moral reform – he died on Friday 3 September 1658. The most likely cause of his death was said to be septicaemia as a result of a urinary infection. Following an elaborate funeral at Westminster Abbey, he was succeeded as Lord Protector by his son Richard.

Unlike his father, Richard had no power base in Parliament, or the army. Weakened by this, he was forced to resign in May 1659, thus ending the Protectorate. With no clear leadership, various factions jostled for power, which eventually saw Charles II emerge from exile in 1660 to restore the monarchy and become king. One of his first acts was to appoint my ancestor William, the man who had plotted so fervently against his father, as Lord Chamberlain of the Household and a Privy Counsellor. Not for nothing was he known for his dexterity as 'Old Subtlety'.

In a fit of revenge, Charles II ordered Cromwell's body to be exhumed from Westminster Abbey on 30 January 1661, the twelfth anniversary of the execution of his father. It was then hung in chains at Tyburn, London, before the decomposed head was cut off and displayed on a pole outside Westminster Hall. Afterwards, it was owned by various people, and was even publicly exhibited several times, before being buried beneath the floor of the antechapel at Sidney Sussex College, Cambridge, in 1960.

As for the New Model Army, Charles II disbanded it shortly after his return. Although it may have died in name, it had emphatically shown what could be achieved by a professional army and it would serve as inspiration all over Europe. While England may have addressed the perilous condition of its army, its navy was soon in a similar state of disarray, and an elite European unit looked to take advantage in one of the most outrageous raids in history ...

15

THE DUTCH MARINE CORPS

AD 1667

As the summer haze wafted over the mouth of the sparkling River Thames, it seemed that England was finally at rest. After two dismal years, which had seen London decimated by plague and fire, England's navy had also suffered damaging defeats in the Second Anglo-Dutch War. But now, with London being rebuilt, the war looked to be at an end and the two countries were engaged in peace negotiations. King Charles II could finally breathe a sigh of relief. However, the Dutch grand pensionary, Johan de Witt, had other ideas. While, on the one hand, he negotiated for peace, on the other he had received information that could allow him to wipe out the entire Royal Navy in one fell swoop. Following a long history of disputes with England, he was increasingly tempted to go for the jugular.

The First Anglo-Dutch War had initially broken out over trade disputes during the early years of Cromwell's Commonwealth. Indeed, I had learnt about the background to some of this as a child living in Cape Town, South Africa. In school, we learnt that British ships had first entered Table Bay in 1601, and, twenty years later, claimed the Cape for King James I. The Dutch, however, arrived thirty years later and ignored this British claim, going on to set up a commercial base. At that time, the land all the way to the south as far as Cape Agulhas was rich in game, which Dutch nomadic 'trekkers' began to decimate. They also killed off great numbers of the desert-dwelling Kalahari San or Bushmen. The Dutch aggressively repeated such actions around the world, leading to great rivalry with England, and eventually conflict.

Although the conflict ended in 1654, with neither side able to claim outright victory, de Witt continued to forge the Netherlands into one of Europe's trading superpowers, while also undertaking a massive shipbuilding programme. Soon the Dutch merchant fleet had grown into the largest in Europe and came to dominate the west African coast, again threatening England's ability to trade. So commenced the Second Anglo-Dutch War in 1665, where two years of bitter fighting followed, with each side inflicting significant blows on the other.

In August 1665, the English fleet had been defeated by the Dutch at the Norwegian port of Bergen, but were soon to suffer far worse. In June 1666, during the so-called Four Days' Battle, the English fleet was decimated by Michiel de Ruyter, one of the most formidable admirals in Dutch history. While the pride of the fleet, the *Prince Royal* had been forced to surrender, the Royal Navy had lost twenty-three further ships, with 1,450 wounded and 1,800 captured. In contrast, the Dutch had lost just four ships and 1,300 wounded.

Whereas the Dutch navy had gone from strength to strength over the previous decade the English navy was in disarray. During the latter years of the Commonwealth and Protectorate, naval debt had rapidly accumulated and funds had dried up, in part due to the upkeep required for Cromwell's New Model Army. Investing little in building new ships, some men went unpaid for up to four years. Soon the arrears in pay owed by the state had rocketed to £400,000. Indeed, by February 1666, the navy had a debt of £2,300,000 and only £1,500,000 available to answer it.

With tickets being issued in lieu of proper pay, it became difficult to recruit seamen voluntarily. Each county was subsequently given a quota to fill but many men went into hiding to avoid service. Even when seamen joined the fleet there was a risk they would desert. Peter Pett, commissioner of Chatham Dockyards, was moved to write to Samuel Pepys, secretary of the Navy Board, about the sorry state of the pressed men. He referred to them as 'those pitiful creatures who are fit for nothing but to fill the ships full of vermin'. The English commander, George Monck, also complained that he had 'never fought with worse officers in his life, not above twenty of them behaving like men'.

It was unsurprising that, with haphazard pay and poor conditions, a number of English subjects defected to the Dutch cause. A few had done so for ideological reasons, but many had been driven by the pursuit of money. The Dutch paid, while the English did not. Many of the English prisoners captured by the Dutch even refused to be repatriated, and offered to fight for them instead. With this, the English navy looked to be no match for the Dutch.

With English pride wounded after the shame of defeat during the Four Days' Battle, it seemed all was lost. However, after a quick rebuilding and recruiting operation, hostilities resumed in July 1666, in the St James's Day Battle. Against all the odds, the English now inflicted a surprising and humiliating defeat on the Dutch, with over 5,000 Dutchmen dead or wounded compared to just 300 English casualties.

But, just as Charles II looked to capitalise on this momentous victory, London was hit by its second major disaster in the space of a year. The city was only just recovering from the deadly bubonic plague, which at its height had led to 7,000 deaths a week and forced thousands to flee safety, rendering London a ghost town. Unsurprisingly, this had done enormous damage to the economy and in turn the navy. Little more than a year later disaster was to strike again when, on 2 September 1666, fire tore through the city.

My father, affectionately known as 'Colonel Lugs', alongside his fellow Royal Scots Greys following a famous victory at Alamein in 1942.

A portrait of my father shortly before he died in 1943. We were never to meet, but I always wanted to follow in his footsteps.

With my special force of Omanis, Baluchis and Zanzibaris and one of our specially equipped patrol vehicles, during the Dhofar Campaign.

The famed Panzer commander Erwin Rommel, also known as 'Desert Fox', finalising battle plans with his men.

Historical knowledge of the Immortals is somewhat limited, save for contemporary depictions such as this, showing them with spears and counterbalances.

While renowned warriors, it was the Immortals' cunning that allowed them to take the town of Babylon in 539 BC.

The infamous Spartan phalanx, which frustrated the superior Persian forces at Thermopylae in 480 BC.

Pelopidas leading the Theban forces at the Battle of Leuctra in 371 BC.

A fragment from the *Alexander Mosaic* (*c.* 100 BC) depicting Alexander the Great in battle against the Persians.

Discovered in Lebanon in 1887, the sarcophagus of Alexander the Great may not contain his remains. It depicts the Macedonian leader during the Battle of Issus in 333 BC.

A Victorian engraving of a frieze of the Praetorian Guard of ancient Rome.

While the Praetorian Guard were meant to protect the Roman emperor, they have become a byword for treachery due to their many attempts to seize power.

Flocking to Europe from Scandinavia, the Varangians, with famed horn helmets and axes, defended and furthered the Byzantine Empire with ferocity and cunning.

The Knights Templar and Hospitallers helping to end the Siege of Acre in 1191.

The Hospitallers' origins might have been charitable, but they evolved to become fearsome warriors during the Crusades.

The great Muslim warrior Saladin leading his Ayyubid forces to victory at the Battle of Hattin in 1187. Pope Urban III died of shock upon hearing of the Crusaders' defeat.

From humble beginnings, Genghis Khan would found the Mongol Empire, which became the largest contiguous empire in history after his death.

In 1221, Subotai led the Mongol Kheshig on one of the most remarkable reconnaissance missions in history, paving the way for the Mongols' invasion of Europe.

From being captured as a slave, Baibars rose to become sultan of the Mamluk Empire, becoming known as the 'Lion of Egypt'.

Their intense training in Cairo meant the Mamluks were renowned for their ability to wield weapons while riding their horses at high speed.

The Janissaries sporting their *börk* hats, which had a holding place in front for a spoon. This symbolised the so-called 'brotherhood of the spoon', which reflected a sense of comradeship among the Janissaries, who ate, slept, fought and died together.

Seen with their arquebusiers, the Janissaries were among the first armies to utilise gunpowder to great effect.

A mercenary army, the Landsknechts were renowned for their infamous pike block formation, aided by a wall of arquebusiers.

Georg von Frundsberg was known as the father of the Landsknechts. It was said he was capable of raising 20,000 men in a matter of weeks.

The Ninja of Iga played a vital role in ushering in the 265-year shogun dynasty, thanks to their ability to wage unconventional warfare.

Oliver Cromwell had no battle experience before the English Civil War erupted in 1642. A brilliant commander, he helped found the New Model Army, leading the Parliamentarian Roundheads to victory.

The New Model Army's newfound professionalism helped Parliament record a decisive victory at the Battle of Naseby in 1645.

The Dutch raid of the Medway in 1667 was one of the greatest humiliations Britain has ever suffered, with its naval fleet being all but wiped out.

Trained at Shorncliffe by Sir JoỸ Moore, the British Light Infantry survived the retreat to Corunna in 1809 and helped finally defeat Napoleon at the Battle of Waterloo in 1815.

Renowned for their unerring bravery, the all-western Iron Brigade suffered heavy losses in helping the Union achieve victory at the Battle of Gettysburg in 1863.

Trained to clear trenches during the First World War, the German Stormtroopers were, for a while, more than a match for British tanks.

RAF pilots seen enjoying a well-earned rest during the relentless Battle of Britain.

RAF pilots in a scramble to their Spitfires as the Luftwaffe invaded.

Approved by Winston Churchill in 1940, the commandos earned a reputation for unconventional raids that helped win the Second World War.

The raid of St Nazaire in 1942 is known as the 'Greatest Raid of All Time'. Here is HMS *Campbeltown*, rammed into the dock, shortly before it exploded.

Masters of deception, the Brandenburgers took the Maikop oilfields in 1942 and helped keep Germany in the war.

Once deployed, paratroopers had to worry not only about a safe landing, but also about being shot out of the sky by the enemy.

British paratroopers on their way to take down the Merville gun battery, paving the way for the D-Day invasion.

On 5 May 1980, the SAS lived up to their motto 'Who Dares Wins' with their successful raid of the Iranian Embassy.

The Green Berets, America's elite special forces, were instrumental in defeating the Taliban during the War on Terror, utilising a mix of state-of-the-art tecŸology, old-fashioned manoeuvres and immense courage.

Helping to avenge 9/11 and take down the Taliban, Staff Sgt Bart Decker of the Green Berets is seen on horseback in Afghanistan.

Founded in 1962 by President Kennedy, the Navy SEALs have become renowned for their tough training regime and ability to wage guerrilla warfare.

With intense training and a fearsome array of weapons, it seems no one is beyond the reach of the Navy SEALs, as Osama bin Laden found to his cost in 2011.

The devastation was tremendous. In four days, an estimated 70,000 people had been made homeless. Over 13,000 houses had been destroyed, as had vital trading buildings such as the Custom House and the Royal Exchange. The cost of the fire was estimated at £10 million, an incredible sum for the time.

Faced with this disaster, there were no longer funds to continue the war. The Dutch were, however, also keen to bring the conflict to an end, with losses mounting up on both sides. With peace now very attractive, the Dutch and English thrashed out terms in Breda. As they did so, Charles II took the step of laying up his entire navy in shipyards, while he also discharged most of the crews of his prize vessels in an effort to save money. Yet, in their absence, all that now protected the laid-up English fleet at the main dockyard at Chatham was a chain across the River Medway. Word of this soon reached de Witt.

The Dutch grand pensionary realised that, with Charles having diverted his attention and financial resources to rebuilding London, he had left his fleet vulnerable to attack. This was a chance not only to destroy the English navy but to hold Charles over a barrel in the peace negotiations. It was a unique but high-risk opportunity, and one that a newly formed Dutch corps could capitalise on.

Two years previously, the noble Lieutenant Colonel Willem Joseph Baron van Ghent, along with de Ruyter, had approached de Witt with an idea. In response to the specially trained English Royal Marines, who had only themselves been set up in 1664, they wanted to create a new regiment of specialist 'ship soldiers'. The regiment would have its own command and serve exclusively at sea, armed with sabres and *snaphaan* guns.

De Witt enthusiastically agreed to the idea and appointed van Ghent as the Dutch Marines' first commander. However, it was an Englishman named Thomas Dolman who was to be put in charge of the 800 troops for the operation on the Thames. Sadly, little is known about Dolman other than he had turned to the Dutch after refusing to accept the Restoration in 1660, having been a friend of Oliver Cromwell and George Monck. While Monck and Dolman had once been friends, soon they would come face to face in conflict.

On the morning of 6 June 1667, a massive Dutch taskforce of sixty-two frigates, fifteen lighter ships and twelve fireships entered the mouth of the River Thames between the Isle of Sheppey and the Isle of Grain. Their arrival caused instant panic. As word of the assault reached nearby Chatham Dockyard, where the majority of the English fleet was based, Peter Pett, the commissioner, immediately raised the alarm. But it was already too little too late; the English were woefully unprepared.

Virtually unopposed, the Dutch Marines made their way to Sheerness Fort, which guarded the approaches to the Medway and Chatham Dockyard. On approach, Captain James van Brakel's ship, the *Vrede*, opened fire on the *Unity*, causing it to withdraw upriver. Meanwhile, Dolman and his 800 marines stormed onto land and charged towards the fort. However, they found that most of the Scottish garrison had already deserted, leaving just seven soldiers behind. Quickly taking the soldiers captive, the marines seized the fort's guns and stores, before pulling down the English flag and hoisting the Dutch flag in its place. The Medway now lay undefended. Unless the English acted quickly, their entire fleet would be lost.

Having been ordered to take charge of the impending disaster, George Monck found things were even worse than he had imagined. Only twelve of the 800 dockyard men expected at Chatham had arrived after the alarm was raised. Moreover, of the thirty sloop sailboats required for defences only ten were available, with the other twenty being used to evacuate the personal possessions of several officials. Incredibly, no munitions or powder were available and no batteries had been ordered to protect the

6-inch-thick iron chain that blocked the Medway. As things stood, Monck had his work cut out.

With the Dutch fleet making its way to the chain, Monck immediately ordered fireships to be sunk before it, hoping to blockade the Dutch. He also ensured that the *Charles V*, *Matthias* and *Monmouth* were moored in defensive positions behind it, while ordering the *Royal Charles*, the flagship of the English fleet, to be moved upriver and out of harm's way. However, Commissioner Pett was unable to find sufficient men to move the ship. They just had to pray the chain held firm.

Yet in the Dutch fleet was a captain with something to prove. Having been arrested for allowing his men to plunder the Isle of Sheppey, Captain James van Brakel begged to be restored to his command, promising to break the chain that crossed the river. It seemed a suicide mission but nonetheless permission was granted. Soon van Brakel's ship, the *Vrede*, along with two fireships, the *Susanna* and the *Pro Patria*, sailed to the front of the Dutch fleet and headed directly for the chain. In response, the English guardships and batteries unleashed a tirade of cannon fire and gunshots, but still the *Vrede* kept going.

Coming alongside the *Unity*, the marines positioned themselves on the mast of the *Vrede* and fired down at the fleeing English soldiers. All the while, van Brakel screamed for cannon after cannon to be fired at close range. In the face of such firepower, the opposition from the *Unity* collapsed and the Dutch marines stormed on board.

While van Brakel took the *Unity*, the Dutch fireship the *Susanna* sailed up to the chain but, after coming under attack, it erupted in flames. Yet, as the English cheered, the *Pro Patria* now sailed through the black smoke and snapped the chain as though it were a piece of string. As she positioned herself alongside the *Matthias*, the marines set her afire, then, with a huge detonation, blew her up. The river to Chatham, and to the English fleet, was now open.

This was the signal the Dutch had been waiting for. Steaming forward, the fleet peppered the *Charles V* with cannon fire and then set her alight. The English crew tried to escape in boats, while others jumped overboard and began swimming to shore, being picked off by the marines' gunfire as they did so.

On seeing this, van Brakel left the *Unity* in a sailboat and made for the burning *Charles V*. Climbing over the bows, with his sword drawn, he ordered a trumpeter to go aloft and haul down the English flag. Another major English ship was now in Dutch hands, although they were unable to put out the fire and instead decided to blow her up.

A Royal Navy historian who witnessed the events stated:

The scene at that moment to be witnessed below Chatham, has not often been paralleled in naval history ... The river was full of moving craft and burning wreckage; the roar of guns was almost continuous; the shrieks of the wounded could be heard even above the noise of battle, the clangour of trumpets, the roll of drums, and the cheers of the Dutch as success after success was won; and above all hung a pall of smoke, illuminated only, as nights closed in, by the gleam of flames on all sides and the flashes of guns and muskets.

After the *Unity* had been taken, the chain broken, and the *Matthias* and *Charles V* set on fire, the Dutch set their eyes on the greatest prize of all: the *Royal Charles*. But, rather than defend the most prized ship in their fleet with their lives, the English chose to jump ship or surrender before any serious fighting got underway. It took just nine Dutch marines to take England's most prestigious ship.

Such was this disaster that Samuel Pepys wrote:

All our hearts do now ake; for the newes is true, that the Dutch have broke the chaine and burned our ships, and particularly

'The Royal Charles', other particulars I know not, but most sad to be sure. And, the truth is, I do fear so much that the whole kingdom is undone, that I do this night resolve to study with my father and wife what to do with the little that I have in money by me.

At this news, the citizens of London fell into a full-scale panic. As rumours spread that the Dutch were in the process of transporting a French army from Dunkirk for an all-out invasion, many wealthy citizens fled the city, taking their most valuable possessions with them. Those who had no choice but to stay prepared to defend their homes with their lives.

Meanwhile, van Ghent's squadron ran aground the *Royal Oak*, *Loyal London* and *Royal James*, with ever more English ships being sunk, set ablaze or captured. In response, Monck could do very little. While he ordered some ships to be towed further upriver, away from the rampaging Dutch, others were deliberately breached to try to slow them down. By the end of the raid, the English had deliberately sunk thirty of their own ships.

Rather than any English defence, it was only the ebbing of the tide that halted the Dutch assault. Pulling away the *Royal Charles* and *Unity* with them, the English sailors who had fought for the Dutch yelled out, 'We did heretofore fight for tickets; now we fight for dollars.'

After one of the greatest humiliations in British naval/military history, coming hot on the heels of the plague and the Great Fire of London, Charles was in no position to defy de Witt. The English navy had lost all but one of its largest warships, which in the short term left the country in a dangerously weak position. He subsequently ordered the peace negotiations to be concluded as quickly as possible and the treaty was signed on 31 July. The terms saw the Dutch hold Surinam, while they were also allowed to import into England goods of their own, or from Germany and south Holland.

There was jubilation in the Dutch Republic after such a famous victory. Many festivities were held in the fleet's honour, while the various admirals were hailed as heroes. De Ruyter and van Ghent were even commemorated on precious enamelled golden chalices depicting the events. Rather than be of any actual use to the Dutch navy, the *Royal Charles*'s draught was found to be too deep for the shallow Dutch waters. It was therefore permanently dry-docked as a tourist attraction until 1672.

Almost 250 years after these events, Rudyard Kipling wrote the poem 'The Dutch in the Medway (1664–72)' for C. R. L. Fletcher's *A School History of England*:

If wars were won by feasting,
Or victory by song,
Or safety found in sleeping sound,
How England would be strong!
But honour and dominion
Are not maintained so.
They're only got by sword and shot,
And this the Dutchmen know!

The moneys that should feed us
You spend on your delight,
How can you then have sailor-men
To aid you in your fight?
Our fish and cheese are rotten,
Which makes the scurvy grow –
We cannot serve you if we starve,
And this the Dutchmen know!

Our ships in every harbour
Be neither whole nor sound,
And, when we seek to mend a leak,

No oakum can be found;
Or, if it is, the caulkers,
And carpenters also,
For lack of pay have gone away,
And this the Dutchmen know!

Mere powder, guns, and bullets,
We scarce can get at all;
Their price was spent in merriment
And revel at Whitehall,
While we in tattered doublets
From ship to ship must row,
Beseeching friends for odds and ends –
And this the Dutchmen know!

No King will heed our warnings,
No Court will pay our claims –
Our King and Court for their disport
Do sell the very Thames!
For, now De Ruyter's topsails
Off naked Chatham show,
We dare not meet him with our fleet –
And this the Dutchmen know!

Though I cannot claim to have destroyed a nation's entire fleet, I did once manage to stop all canal traffic across Europe. While on a canoeing operation with the Scots Greys in the Schlei River, one of my corporals accidentally fired a flare onto the deck of a huge Russian tanker. Our flare hissed and burned away fiercely on the boat deck and in a short while the klaxon and red-light system, which was installed along the Kiel Canal banks right across Europe, began to honk and flash as though World War Three was about to erupt. Loudspeakers crackled and a disembodied male voice spoke to us with British Rail-like intelligibility. I understood only two words with crystal clarity: *'Englander Soldaten'*. A British-style beret or cap comforter must have been spotted on a canoeist. At this, I ordered the immediate end to our exercise and ensured we scarpered.

It soon transpired that, because of the flare, all canal traffic across Europe had stopped for five hours, which was an expensive delay. Six months later a forester found a Grey's beret by the canal and handed it to the police. I was immediately summoned to see my CO. Realising that my team and I had been responsible for the flare on the tanker, he asked if I had any idea of the consequences if the highly flammable tanker had exploded? While I tried to play dumb, I was assured that an international incident might have occurred, and the Cold War might have suddenly become very hot. For this, I received a hefty fine.

The Dutch raid was an almighty wake-up call for the English navy. It now had no option but to rebuild the fleet with newer and larger ships. In time, this would establish the Royal Navy as a major fighting force for the eighteenth century, a position that led to it securing command of the world's oceans. Yet the rise of Napoleon soon represented a new threat. If England was to prevail, it once more had to face abject humiliation before a new elite force of soldiers would deliver victory ...

16

THE BRITISH
LIGHT INFANTRY

AD 1809

Like a scene from the Old Testament, thousands of bedraggled, emaciated figures emerged from the snow-covered mountains and staggered towards the Spanish port of Corunna. With their matted beards, bloodied bare feet and stale stench, these men could easily have been mistaken for a horde of beggars, if not for one thing – their filthy, torn, red uniforms. For this was the might of the British Army and, if they didn't get home soon, they might all be wiped out.

For weeks, they had fled from Napoleon's forces, retreating for over 250 miles, in perilous conditions, in order to reach the ocean. Here, their leader, Sir John Moore, expected a convoy of ships to be waiting. However, as they reached the port, they found it was empty. There appeared to be no way to get home, while Napoleon's men were advancing hard on the rear. If they were still going to be alive by the time the ships did finally arrive, then they would have to fight. And while England's armed forces looked to have nothing left, Moore had one card left to play: the British Light Infantry. To prevent total obliteration, they were the British Army's only hope.

Yet, without the inspiration of the French, and the genius of Sir John Moore, it was unlikely that they would have even been formed.

The French were experts at utilising their light infantries, with their so-called chasseurs, helping Napoleon to a number of victories ever since he had seized power in 1799. The British, meanwhile, had been left behind. Despite these units offering greater mobility on the battlefield, British attempts to embrace the concept had been half-hearted at best. Unbelievably, some British commanders were resistant because they believed that fighting from cover was in some respect dishonourable. Yet, if Britain was to prevail against Napoleon and his chasseurs, it had to take building a light infantry force seriously. The man it looked to achieve this was Sir John Moore.

The son of a Scottish doctor, Moore was born in 1761 and, after joining the army in 1776, had risen to the rank of lieutenant general of the 52nd Regiment. His knowledge of military tactics, and his good relations with his troops, were said to be second to none. However, it was his experience with light infantry that made him particularly ideal for the role, having utilised such a corps when serving as military governor of St Lucia.

As such, in 1803, following the outbreak of war between Britain and France, Moore was appointed commander of a new brigade based at Shorncliffe Army Camp in Kent. This brigade was designed to serve as the basis of the permanent light infantry force, with Moore offering his own regiment of line infantry, the 52nd Oxfordshire, for training. Moore subsequently chose fellow Scot, Colonel Kenneth Mackenzie, to take command.

At Shorncliffe, the teaching of musketry was considered of the first importance. From taking immaculate care of their arms to efficiently loading them, the troops incessantly practised shooting at targets from a range of distances, at rest and on the move. A firing book was even kept by the captains so that they could record the men's progress.

Speed was also vital to light infantry so drills were practised to the point of tedium, with each soldier knowing the various calls verbatim. Of utmost importance was the so-called 'firing in extended order', whereby, fighting in pairs, one would shoot while the other would reload, ensuring there was virtually no break between them. Skirmishers also practised 'chain order', which meant that, when the first shot was fired, one man from each group fired at a time, so that continuous fire was maintained.

Rearguard drills were also crucial, with the object being not to fight but to delay. It was also understood that, should a rearguard be pursued by an enemy, it would divide itself into two bodies, with skirmishers sent out to protect them. As the skirmishers would need to move quickly, and sometimes react without any direct orders, the men were encouraged to develop initiative and self-direction. These drills would in time prove to be vital to the survival of the British Army.

However, perhaps more than anything at Shorncliffe, instilling discipline was top of the agenda. Colonel J. F. C. Fuller, in his book *Sir John Moore's System of Training*, said of this:

> We may look in vain for a written system of discipline that made the Light Division famous; for no true system of discipline can be set forth within the covers of a book. It is the ceaseless minutiae of daily work, the actions, care and supervision by the officers, the willingness of the men, the interest in the work, the perfecting little by little, and above all, the self-reliance and good comradeship of all ranks that go to build up discipline, esprit de corps and efficiency; not the rules of well-intentioned pedants and the regulations of learned doctrinaires.

Instilling discipline might sound draconian but Moore's warm character, and magnetic personality, ensured that was never the case. Sir Henry Banbury, a friend of Moore's, said of him: 'I knew him well, and loved him, he was always kind to me ... He was thoroughly honourable, just and generous; far above all sordid motives and hardly to be swayed by any passion from what he felt to be his duty.' Sir William Napier, whose brother Charles was trained by Moore, echoed these thoughts: 'His was the fire that warmed the coldest nature.'

As Moore prepared his light infantry in Shorncliffe, the Napoleonic Wars continued to rage across the continent. Napoleon had already forced the Austrian, Russian and Prussian Empires out of the war, and, by 1808, he had set his sights on Portugal and Spain. However, while British forces were successful in forcing him from Portugal, the French moved into Spanish territory, occupied Madrid and installed Napoleon's brother, Joseph, as king. While this provoked an explosion of popular rebellions across Spain, no organised attack against the French came. With the fall of the monarchy, constitutional power had devolved to local juntas, who engaged in a haphazard resistance at best.

Many in Spain now looked to Britain as its last hope. However, scandal had erupted following the victory in Portugal. British commanders, such as the Duke of Wellington, had been called home to answer questions about why they had allowed French soldiers to evacuate instead of being compelled to surrender. In their absence, Sir John Moore was sent to Portugal to command 30,000 troops.

At the same time, Napoleon himself was en route to Spain. Enraged at the loss in Portugal, he was now concerned that his brother might be forced out of Madrid, with insurrection growing throughout the country. He determined that his commanders were to blame and decided to take personal control of the situation, intending to conquer Spain in just two months with a fresh influx of troops.

To face off this threat, Moore was ordered to take 20,000 men and advance into Spain. This was not just the largest army that

Britain had in the field, it was virtually the only army available. As such, some 15,000 reinforcements, under General David Baird, were also at sea and on their way to Corunna, a port in the north-western corner of Spain. Following behind were Moore's light infantry divisions, including the 1st Battalion 52nd, 1st Battalion 43rd and 1st Battalion 95th, fully trained and now under the command of Major General Robert Craufurd.

Moore had every right to believe that with this influx of troops, and the addition of his light infantry division, he could before long hold Madrid and force Napoleon from Spain. Yet the operation soon turned into a disaster. With Moore and Baird both facing troubles on their journey to Madrid, including poor roads, untrustworthy guides and dreadful weather, their effort to free the capital was doomed to failure. Napoleon had over 250,000 troops at his disposal and had retaken Madrid before Moore and his men even arrived. Moreover, while Moore was unaware of this, Napoleon knew that the British were on the march and was prepared for them.

When British forces engaged with the French in bitterly cold conditions, near the town of Sahagun, the half-frozen troopers had trouble holding their reins and sabres with their numbed fingers, while their horses lost their footing on the ice. Despite managing to hold off the French, Moore learnt that Napoleon, and the main bulk of his army, were now heading their way, aiming to destroy Britain's only field army. This would be a catastrophic blow that would surely put the British out of the war and at the mercy of Napoleon.

Moore decided he had no option but to retreat to the port of Corunna and get his men home. But the journey would not be easy. They would need to travel over 250 miles of frigid mountain roads, with the French chasing them all the way. In such circumstances, he entrusted the safety of the British Army to the light divisions that he had once trained at Shorncliffe. They would act as the rearguard, fighting off and delaying Napoleon's forces, keeping the bulk of the army safe until they reached Corunna.

So began one of the most arduous retreats in British military history. In appalling weather conditions, heavily laden commissary and baggage wagons were pulled up tortuous mountain tracks by exhausted oxen and mules, while the men went days without sleep. Moore was determined to reach Corunna before it was too late, but such were the conditions that some men collapsed and died in the snow, their red jackets not enough to keep out the cold. For those who managed to keep going, many were forced to walk barefoot after their boots disintegrated, and thus suffered frostbite. Anyone who tried to prop up an ailing comrade ran the risk of being left behind. Rain, heavy snow and sleet alternated, soaking every soldier through to the bone. With food at a premium, their gnawing hunger was also growing ever stronger.

Such conditions are liable to make even the calmest man snap. I remember when, on an expedition in Antarctica, my team and I were trekking in the dark, up to our knees in snow, battling against a windchill factor of minus 120. It was so cold that the natural liquid in our eyes kept congealing. In such circumstances, it was very difficult to avoid outbursts of temper at each other and the smallest thing would trigger hours of silent hostility. Explorer Ashley Cherry-Garrard said of Captain Scott's winter party during its failed trip to the South Pole: 'The greatest friends were so much on one another's nerves that they did not speak for days for fear of quarrelling.' I know how they felt. Once my partner Mike Stroud somehow managed to break our only two vacuum flasks, which meant we could no longer have a hot drink. Mike's solution was that we should instead drink our tea, or pre-stress energy drink, from the communal pee-bottle. If I wasn't already annoyed at Mike for breaking our two flasks, the prospect of drinking from a bottle that had previously contained urine tested me to my limit. I remember saying, 'It's one short step from cannibalism.' However, I must confess that I noticed no change in the taste of the pre-stress, which was foul at the best of times. On the British Army's march to Corunna, I can only imagine many of these men being at each other's throats, without the necessary clothing, equipment and food they needed for such an ordeal, making matters worse.

Indeed, trekking for hundreds of miles in wet and freezing conditions can also put a terrible strain on your body. On one occasion, my feet were so cold, and red raw, that despite my best efforts to dress them, when I removed the dressing at the end of the day, the outer half of my little toe, along with a chunk of flesh, came away, right down to the bone. I had no choice but to continue onwards, and limping every step of the hundreds of miles still in front of me was agonising and certainly tested my patience. The combination of slow starvation and massive energy expenditure also takes its toll. During our 1,700-mile trek across the Antarctic, where Mike and I pulled sledges weighing 485lbs, 16 miles a day for ninety-three days straight, tests revealed some alarming results. Our blood samples showed that our whole enzyme systems, everything that controlled our absorption of fat, were changing and we were recording levels of gut hormones twice as high as were previously known to science. Furthermore, with zero remaining body fat, we were losing muscle and weight from our hearts as well as our body mass. In similar conditions, it is no wonder that so many soldiers on the trek to Corunna simply sat down and died in the snow.

In the rearguard, things were thankfully a little better. With the 1/52nd and 1/95th battalions, commanded by Edward Paget, and the 1st Flank Brigade, now comprising the 1/43rd, 2/52nd and 2/95th commanded by Robert Craufurd, the light divisions displayed all of the discipline and skill that Moore had taught them at Shorncliffe. Skirmishing in pairs, one would fire their musket while the other would load. Darting in and out of rocks on the mountains they deployed themselves in high positions and were a constant pest to French cavalry trying to advance.

From their position guarding the crossing at the Esla River, two privates of the 43rd, John Walton and Richard Jackson, suddenly saw the French advance guard approaching. As Jackson raced to inform Craufurd of this, the French cavalry slashed at him, inflicting up to fourteen sabre cuts. Refusing to go down, he managed to stagger away and raise the alarm. Meanwhile, Walton was left to defend the bridge by himself. Shooting and diving for cover to reload, his clothing was shredded by sabres and his bayonet bent double. Despite this, he somehow managed to keep the French at bay long enough for Craufurd to arrive.

Soon the Light Brigade positioned themselves in formation on the banks of the river. Benjamin Harris of the 95th recalled that the rain was so heavy that it actually flowed out of the muzzles of their rifles. Upon seeing the light infantry in position, and having been forced back by Walton alone, the French declined to engage, instead waiting for their main army to join them. Craufurd could not wait for this to happen. Engineer Captain J. F. Burgoyne was subsequently deployed to place explosives on the bridge and detonate it. As he did so, Napoleon and his army arrived but it was too late. Burgoyne set off the explosives and the bridge was blown sky-high. Napoleon could only look on as the British rearguard escaped, having bought their army more time to reach Corunna.

While the Light Brigade covered itself in glory, the main British forces did anything but. Forced by heavy snow to stop at the town of Astorga, British and Spanish soldiers, supposedly allies, fought for the best billets, while hordes of redcoats prowled the streets in search of liquor, looting shops and houses with inebriated abandon. The British Army was fast disintegrating and discipline was washed away in a sea of wine and rum. When he saw this, Moore doubted they would make it to

Corunna. His men were clearly in no fit state to fight Napoleon should they be caught. As such, he ordered that ammunition and other stores be blown up. Not only were such items weighing them down, but Moore would rather put the wagons to the torch than have them fall into the hands of the enemy. Later, Moore would also be forced to throw casks containing $25,000 over the mountains for the very same reason. Yet, while the British Army was floundering, the rearguard of the Light Brigade dealt another significant blow.

With French forces pouring forward, the Light Brigade was forced to fall back to Benavente. In doing so, they were hotly pursued by 600 of Napoleon's infamous chasseurs of the imperial guard. Believed to be the finest horsemen in Europe, and led by one of Napoleon's favourite generals, Lefebvre-Desnouettes, they looked to impress their master by wiping out the British rearguard.

However, Paget was aware of this, and had set a trap. With their blood racing, the chasseurs became less cautious, piling towards Benavente, where Paget's 10th Hussars were hiding, lying in wait. Before the chasseurs knew what was happening, the trap was sprung. Paget's men crashed into the surprised Frenchmen at the gallop; steel clashed with steel, with the British swords so sharp that French heads and limbs were severed at a single stroke, while the light infantry let off a flurry of shots.

The chasseurs broke, starting a 2-mile running fight back to the river. Forced to retreat, French troopers plunged into the icy water, hoping to swim to safety. In their heavy uniforms, many drowned. The light infantry also claimed a notable scalp, wounding Lefebvre-Desnouettes and taking him prisoner. Some seventy-five chasseurs joined him in captivity, while another fifty-five lay dead or wounded on the field. In contrast, British casualties numbered around fifty.

Napoleon was rocked. His favourite general had been taken prisoner while somehow the British rearguard had managed to keep his troops from progressing. All the while, the bulk of Britain's armed forces was slowly but surely reaching Corunna. But now the time had come to focus on preserving one of Britain's elite units. Virtually all of its light infantry was fighting on the flanks and rearguard. While they had achieved exceptional results, sooner or later, Napoleon's men would surely overwhelm them and wipe them out. Moore therefore ordered Craufurd's 1st Flank Brigade to break off and head to the port of Vigo. They would not only help to guard the main body's southern flank as they did so, and hopefully encourage some of the French forces to splinter, but would also ensure the survival of the light infantry.

Although it was only the road and the elements against which they had to battle, their subsequent march over rugged and frozen terrain became a trial of the greatest intensity. That they reached Vigo in some order, and not as a mob of stragglers, was due partly to the quality of the troops and partly to the efforts of Craufurd. Driving them on, he kept them together by enforcing the strictest discipline. When a rifleman named Howans had grumbled within earshot, Craufurd sentenced him to 300 lashes. Craufurd's leadership skills, including the meting out of such punishment, were celebrated by Benjamin Harris of the 95th in his memoirs, *The Recollections of Rifleman Harris*:

No man but one formed of stuff like General Craufurd could have saved the brigade from perishing altogether; and, if he flogged two, he saved hundreds from death by his management ... He seemed an iron man: nothing daunted him – nothing turned him from his purpose. War was his very element, and toil and danger seemed to call forth only an increasing determination to surmount them ... I shall never forget Craufurd if I live to a hundred years I think. He was everything a soldier.

They staggered on until they reached Vigo on 12 January 1809, when the sight of the sea revived their spirits, as William Surtees recalled in his *Twenty-Five Years in the Rifle Brigade*: 'Fellows without a shoe or a stocking, and who before were shuffling along with sore and lacerated feet like so many lame ducks, now made an attempt to dance for joy.' Craufurd and his men finally reached England on 27 January, ensuring the light infantry would at least survive, if nothing else.

With Craufurd and his men heading to Vigo, and home, Moore placed the responsibility of maintaining the rearguard on Edward Paget's division, plus the 15th Hussars. Meanwhile, the British forces continued to degenerate as cold, hunger and insanity took hold. As they marched through towns, Spanish civilians often fled from the British, who now resembled a barbarian horde more than a modern army. Those villagers who did stay were often robbed, mistreated and at times even murdered, as mobs of drunken soldiers sacked towns such as Villafranca. Moore desperately tried to restore some order by giving a heartfelt speech to his ragged redcoats, and appealed to their sense of duty, honour and allegiance to their country. To reinforce his words, a looter was hanged in front of the assembled troops. But it was no use – the marauding continued. Dismayed, Moore said to his troops, 'Soldiers, if you do not behave better I would rather be a shoeblack than your general.'

Those who refused to obey orders, and hid in the wine cellars, lying sprawled about in a wine-induced stupor, soon paid the ultimate price. Being left behind, over 1,000 of them were caught by the French cavalry and ruthlessly cut down.

Back at the rearguard, fighting had broken out over the bridge crossing the River Cua at Cacabelos. Leading the French advance guard, General Colbert charged forward, before coming under heavy fire from the light infantry divisions concealed behind stone walls across the bridge. Shot after shot was fired in the pouring rain, suddenly halting the advancing French, causing one British participant to remark, 'I never saw men ride more handsomely to destruction, until we poured it into them right and left, and they went down like clockwork.'

Such was the light infantry onslaught against the French that according to ensign Robert Blakeney of the 28th the road was 'absolutely choked with their dead'. Among them was General Colbert. As the general rode across the bridge, waving his sabre, Paget looked to Rifleman Thomas Plunkett, his most deadly marksman. Offering him a reward if he were to shoot Colbert dead, Plunkett emerged from behind a wall, lifted his Baker rifle and, from a distance of 400 metres, shot the French general in the head. As a trumpet major rushed to Colbert's aid, Plunkett reloaded, raced forward and shot him dead too, proving the first shot to be no fluke. At this, any Frenchmen who had thought of crossing the bridge fell into retreat.

Thanks to the monumental effort of the rearguard, the British Army finally reached Corunna on 11 January 1809. Despite their shocking appearance, they were overjoyed to have made it. In all, Moore had lost around 5,000 men during the retreat, and another 3,500 had taken ship at Vigo. The sight of the sea caused some of the men to shout as though they had 'beheld a deity'. The weather had also improved and was now almost springlike. But there was a problem. The ships Moore had been counting on to take them home had still not arrived. Until they did so they faced an agonising wait, with Napoleon's forces getting ever closer, despite the best efforts of the rearguard, who blew up the bridge at El Burgo, just 4 miles from Corunna, before retreating to the town itself.

With the ships three days away, Moore had no choice but to prepare his men, and the town, for war. Re-equipping his soldiers with new muskets from stores, he also proceeded to destroy anything that could be of use to the French, including 12,000 barrels of gunpowder and 300,000 musket cartridges. The resulting explosion was so enormous that it caused some structural damage to the city.

Finally, the 100 transports and twelve warships arrived on 14 January to much cheering, with the French forces still having not reached the town. Frantic loading of the ships commenced, with any horses not fit enough for the journey having to be shot. But, just before they were ready to get away, Moore received word that the French were just 2 miles away and advancing fast. They needed to be stopped.

Moore quickly placed a third of his army, predominantly infantry with a few guns and cavalry in support, to defend the Monte Mero, a low ridge 2 miles south of Corunna. While the British troops had looked close to death moments before, the sound of French gunshots saw them suddenly find reserves of energy, now that the ships had arrived and home was so close. Representative of the general spirit of the army was an officer of the 32nd. He was so exhausted by the retreat that he was scarcely able to stand, so his men found him an armchair in which he sat during the battle, so as not to miss the fighting.

Yet, despite their best efforts, the British soon looked to be overwhelmed. As the 42nd were pushed back, Moore exhorted: 'My brave Highlanders! Remember Egypt!' This brought renewed vigour. In response, the battalion leapt up, cheered and charged forward. Watching proudly from his saddle, Moore was suddenly wrenched from his horse. He had been shot, his left arm blown away at the shoulder. Quickly placed in a blanket, he was carried towards Corunna by men of the 42nd.

While Moore clung to life, the light infantry, whom he had raised and trained at Shorncliffe, did him proud. The 1/95th and 1/52nd fell into skirmish order, raking their way across the mountains, displaying lethal marksmanship, until they ran out of ammunition. Their efforts saw them take seven officers, and 156 other ranks, prisoner. More importantly, they had managed to hold off the French until darkness fell so the British could now board their ships. But Moore was not to join them, having passed away shortly before. However, he had lived long enough to know that the French had been defeated and his army was safe. Just before he passed, he said, 'I hope the people of England will be satisfied. I hope my country will do me justice.'

The fleet carried home 26,000 men of Moore's army, many of whom would fight again and in time return to the continent to play a key role in helping Britain win the Peninsular War in 1814. This proved a major disaster for the French, with France and her allies losing at least 91,000 men and 237,000 wounded. But the real decisive blow would land a year later at the Battle of Waterloo, where the Duke of Wellington was joined by Moore's 1/52nd. Far from an insignificant force, as in the past, this light infantry division was now the largest battalion at Waterloo, numbering 1,130 men.

On 18 June 1815, as Napoleon looked to be on the verge of a famous victory, with his Middle Guard launching an assault on the British line, the 52nd charged forward and fired volley after volley against their left flank. William Hay, a Light Dragoon watching from the right, later recalled that 'so well-directed a fire was poured in, that down the bank the Frenchmen fell and, I may say, the battle of Waterloo was gained'. With the 52nd rapidly advancing, they forced Napoleon's guard into full retreat. According to Wellington, the battle was 'the nearest-run thing you ever saw in your life'. With the Battle of Waterloo won, thanks to the efforts of the 52nd, and of course the Royal Scots Greys, Napoleon was finally vanquished. Exiled to the remote island of Saint Helena in the South Atlantic, he died six years later at the age of fifty-one.

On the fifty-third anniversary of Waterloo, Colonel George Gawler of the 52nd Light Infantry, wrote to his son, claiming that victory was down to Moore's training at Shorncliffe:

This really wonderful, thoroughly fought-out battle was won under God (by British sturdiness no doubt), but under the drill system of Pliable Solidarity. Stiff Solidarity characterised the European Armies up to the French Revolution of 1792. Then the wild sans-culottes, the French, were obliged to assemble, and adopted the system of elan with as little of the solidarity as they could do with. Then good common-sense heads in England devised, first under Sir John Moore at Shorncliffe, Pliable Solidarity. With this system, the old Duke outmanoeuvred every army opposed to him, and never lost a battle. To the very end of the day (Waterloo), we manoeuvred by well-formed battalions, as smoothly and as rapidly as we should have done on Southsea Common. While, from the beginning of the day, French elan, like soda-water, had to be corked up in masses. The moment the density was rudely broken, all went off in smoke and confusion.

While some had severely criticised Moore for his retreat at Corunna, it had allowed the bulk of the British Army, and its light infantry divisions, to survive, regroup and claim victory for another day. The historian Sir John Fortescue has said of Moore: 'If not a stone had been raised nor a line written, his work would still remain with us; for no man, not Cromwell, nor Marlborough, nor Wellington, has set so strong a mark for good upon the British Army as John Moore.'

Light infantry divisions had now more than proved their worth on the battlefield, and would continue to do so, particularly in the American Civil War. Yet, while weapons and tactics would continue to revolutionise the battlefield, there is sometimes no substitute for courage beyond all reason. And one elite division in America would go down in legend for their ferocious bravery against all odds ...

17

THE IRON BRIGADE

In the wake of President Abraham Lincoln's attempt to ban slavery in 1860, war had erupted between the pro-slavery Confederate states and anti-slavery Union states. Despite an initial string of victories for the Union, in 1862, General Lee's Confederate forces swept into Kentucky, crossed the Potomac into Maryland and made for Frederick. Suddenly, the end of the Union seemed to be a distinct possibility.

If the Union was to sustain the war in the east, it had to halt Lee's progress by any means necessary. This grave responsibility fell to General George B. McClellan. Despite the odds being stacked against him, he had something vital in his favour. McClellan's men had seized an order that detailed Confederate plans and locations. With this, McClellan now believed he had the upper hand and looked to target Lee's forces in the area of Boonsboro-Hagerstown. But this wouldn't be easy. His Union forces would first need access to the Potomac through three gaps in the South Mountain: Fox's Gap, Turner's Gap and Crampton's Gap. If they could succeed in taking them, then they would have the Confederates on the run.

One of the men to whom McClellan turned to undertake this vital operation was John Gibbon and his all-western unit. Just a year previously, this ragtag group had answered the call by Governor Alexander Williams Randall of Wisconsin to form an all-Wisconsin unit to help the Union war effort. Soon after, the 2nd, 6th and 7th Wisconsin units, as well as the 19th Indiana Volunteer Infantry Regiment, arrived in Washington to report for their three-year term of duty.

Under their initial commander, Brigadier General Rufus King, the men were at first assigned to defend the Capitol. However, upon Captain John Gibbon taking command in May 1862, and the unit being joined by his Battery B of the 4th US Artillery, the men from Wisconsin were soon thrust into some of the major battles of the civil war.

Having graduated from West Point Military Academy in 1847, Gibbon had seen duty in the war with Mexico as well as against the Seminole Indians in Florida. Following this, he returned to the US Military Academy to serve as an artillery instructor. However, at the commencement of the civil war, despite his North Carolina, Confederate, roots, he was promoted to captain and joined the Union's Battery B, while his three brothers joined the Confederates.

When he took command of the Wisconsin unit, Gibbon was not impressed. They were surly, unruly and their uniforms a ragged mismatch. In short, they appeared to be unfit for battle. Much like Sir John Moore before him, Gibbon's first objective was to tighten discipline. This was not greeted with any great enthusiasm and his initial efforts did not win favour in the ranks. One Wisconsin private reported: 'We are reduced to strict military subordination, and Gen. Gibbon is bound to make regulars of us.' While another said of Gibbon, 'Probably no brigade commander was more cordially hated by his men.' And yet it seems that Gibbon's attempts to instil discipline were not that exacting.

While each soldier was ordered to take a bath once a week, there was also a daily review at 5 a.m., with everyone expected to drink a cup of hot coffee immediately after, whether they liked it or not. In addition, regimental officers who had not been attending the early morning roll call were now required to do so.

Gibbon also sought to address the uneven appearance of his regiments. All soldiers in the unit were now issued with a nine-button dark-blue frock coat, along with white linen leggings and white cotton gloves. However, it was the tall model 1858 Hardee dress hat for which the unit would soon be known, making the westerners stand out on the battlefield.

While many of Gibbon's changes had been met with hostility, the new uniforms were enthusiastically received. 'We have a full blue suit, a fine black hat nicely trimmed with bugle and plate and ostrich feathers,' a 7th Wisconsin man wrote home, 'and you can only distinguish our boys from the regulars, by their [our] good looks.'

Yet, rather than for good looks, this group of ragtag soldiers were to become renowned for their unstinting bravery. Upon meeting Confederate forces at Gainesville, in a battle widely known as Brawner's Farm, the subsequent open-field fighting was brutal. With nowhere to hide, it was a shootout to the finish. Major Rufus Dawes of the 6th Wisconsins wrote of the battle in his memoir: 'Our one night's experience at Gainesville had eradicated our yearning for a fight. In our future history, we will also be found ready but never again anxious.' The casualties were enormous. Refusing to leave the field, more than one-third of the brigade – 725 men – were killed. To illustrate this courage, Lieutenant Colonel Charles Apthorpe Hamilton was shot through the thighs but maintained in the saddle, while his boots filled with blood. Likewise, Major Thomas Allen was shot in the neck and left arm, but continued to fight.

The bravery of the Wisconsins in such circumstances astounds me. While on duty in Oman, the prospect of being killed or wounded was one that kept me up at night. Of particular terror was the thought of some bullet, shrapnel shard or mine blast ripping out my genitals or blinding me. I had seen it happen all too often and it was only by the grace of God that the only injury I picked up in Oman was a crooked finger. Yet these terror images totally dominated my mind. To cope, I learnt to keep a ruthlessly tight clamp on my imagination. With fear, you must prevent, not cure. Fear must not be let in in the first place. If you are in a canoe, never listen to the roar of the rapid ahead before you let go of the river bank. Just do it! Keep your eyes closed and let go. If the fear then rushes at you, it will not be able to get a grip, because your mind will by then be focusing on the technical matter of survival. I wonder if any of the Wisconsin men felt like I did and also had to use some mind trick to blank out fear and charge forward nonetheless?

However, despite the unit's daring deeds, there had been no one but themselves to witness their heroics. The Battle of South Mountain would change all that and spawn a legend. As, while Gibbon's Wisconsin units set off believing they had the upper hand over the Confederates, they had no idea that General Lee had not only reinforced the South Mountain defence, but was also lying in wait.

Arriving at the mountain around 3.30 p.m., Gibbon and his men found that rising sharply on both sides were wooded hills, as well as stone-wall barriers, ravines and farm buildings. Through it all was their target, Turner's Gap, a narrow defile through which the National Road passed. There was only one way in, and one way out. It was vital they captured it.

As the sun began to set behind the mountain, producing an orange haze, Gibbon mounted his horse. It was time to move. Riding to the front line on the high ground, from which he could view his whole force, Gibbon suddenly shouted, 'Forward! Forward! Forward!'

At this, his men charged up the mountain, expecting to meet moderate defences at best. But suddenly there was the sound of rifle shots booming through the air. They were being ambushed.

The Confederates had been lying in wait all this time and now emerged from behind logs, fences, rocks and bushes.

With shells exploding around them, and men collapsing to the floor in pools of blood, the Wisconsins refused to retreat. Moving steadily up the steep mountainside, they unleashed a barrage of gunfire on the Confederate hiding places only to suddenly come under attack from the woodlands. The Wisconsin units now stopped in their tracks. They had not been expecting this. Out in the open, they would all die if this continued, but they could not move forward until the woodland was clear. Yet this was a suicide mission that would involve charging across an open field towards it. It mattered not. As a handful charged headlong towards the woodland, all guns blazing, others threw several shells into a farmhouse, from where gunfire had also been coming. This bought the rest of the brigade vital time to move forward.

In the face of increasing rifle fire and heavier casualties, the brigade steadily advanced for another three-quarters of a mile, driving the Confederate outposts into the narrowing gorge. Positioning themselves behind a stone fence, the Confederates poured hot fire into the advancing Wisconsins. One after another was shot down, almost as if being subjected to a firing squad. Yet still the Wisconsins continued, ever relentless, refusing to be beaten.

Charging out front, right and left, they suddenly outflanked the fence and sent the Confederates fleeing. But, from stone walls and hills 40 yards away, a thousand rifles fired down from both sides, lighting up the gathering darkness and halting the westerners' advance.

Although exhausted from the climb and the ceaseless fighting, the massed men now rushed forward, shouting wildly from throats burning with powder and smoke. As the battle roared towards its climax, flashes of gunfire lit up all sides of the mountain. Such was the unrelenting ferocity of the attack, the Wisconsins' rifles became too hot to reload, and too full of carbon for safe use. Some even ran out of ammunition. At this critical moment, locked in the dark with a powerful enemy, it appeared that the brigade, for all its hard fighting and heavy losses, would have to surrender. But this was far from Gibbon's mind. With no guns left, Gibbon turned to his men and shouted, 'Hold your ground with the bayonet.'

With their bayonets at their side, the bedraggled and bloodied Wisconsins charged into the advancing Confederates. Man after man was shot down but those who reached the Confederate line did so with such savage intensity that they forced the enemy back to a stone barricade. Stunned by this assault, the Confederates found that they had also run out of ammunition. General Lee now had no option but to order an immediate withdrawal.

Up and down the battered western line, word was passed that the battle had been won. In the smoky darkness, three cheers went up for the 'Badger State'. With victory sealed, and Turner's Gap captured, they now had to turn their attention to helping the many wounded and dying men scattered up and down the mountainside. Rufus Dawes was engaged in this terrible task:

> Several dying men were pleading piteously for water, of which there was not a drop ... nor was there any liquor. Captain Kellog and I searched in vain for a swallow for one noble fellow who was dying in great agony from a wound in his bowels. He recognised us and appreciated our efforts, but was unable to speak. The dread reality of war was before us in this frightful death, upon the cold hard stones. The mortal suffering, the fruitless struggle to send a parting message to a far-off home, and the final release in death, all enacted in the darkness ...

Over 25 per cent of Gibbon's men had been lost in the attack. However, the manner in which those losses had occurred had not gone unnoticed. McClellan, as well as a number of other officers, had seen it all – the initial advance up the mountain-side, the unflinching progress as the enemy's fire increased, and the dogged movement, always forward, into the darkness, marked towards the end by the flashes from the opposing lines of rifles.

Apparently, as General Joseph Hooker of the 1st Corps met with General McClellan to get further orders, McClellan asked, 'Who are those men fighting in the Pike?' Hooker replied that it was the western brigade of General Gibbon, to which McClellan answered, 'They must be made of iron.' From here on, Gibbon's men would be known as the 'Iron Brigade'. A few days later a report in the *Cincinnati Daily Commercial* ensured that this name was known far and wide:

> The last terrible battle has reduced this brigade to a mere skeleton; there being scarcely enough members to form half a regiment. The 2nd Wisconsin, which but a few weeks since, numbered over 900 men, can now muster but fifty-nine. This brigade has done some of the hardest and best fighting in the service. It has been justly termed the Iron Brigade of the West.

At Fredericksburg in December 1862, and again at Chancellorsville in May of 1863, the Iron Brigade continued to earn plaudits for its unflinching bravery in the face of overwhelming enemy supe-riority. However, it was the Battle of Gettysburg between 1 and 3 July 1863 that really cemented the reputation of the Iron Brigade as one of the bravest military units of all time.

Under the command of General Solomon Meredith, the Iron Brigade was now the 1st Brigade of the 1st Division of I Corps of the army, as well as being custodians of the flag of the 1st Division. One proud Black Hat proclaimed that this identity was 'purchased with blood and held most sacred'.

As General Lee's forces moved east, to seek a store of shoes in the little town of Gettysburg, Pennsylvania, the Union marched towards them, with the Iron Brigade leading the way. Despite the fact that the enemy appeared to have superior numbers, the Iron Brigade hurled themselves into action. Soon they were in a head-on collision with the enemy, as one eyewitness recorded 'a line of ragged dirty blue crashed into one of dirty, ragged butternut'.

The Iron Brigade swept the Confederate soldiers back. Temporarily overwhelmed, the Confederates started to surrender or fall back to Willoughby Run. The first phase of the battle west of Gettysburg had been won by the Union, in no small part due to the Iron Brigade, but more was to come.

Following a lull of more than two hours, the Confederates attacked again at 3 p.m. with a barrage of artillery fire. The Iron Brigade fired back such a thunderous response that, for a time, 'no rebel crossed the stream alive'. However, Solomon Meredith became a casualty at this stage, having been crushed beneath his horse. While the Union front remained impregnable, the flanks began to yield, forcing them to pull back to a barricade of rails on Seminary Ridge.

From behind this feeble barricade, the Iron Brigade stemmed the fierce tide that pressed upon them incessantly, and held back the enemy lines, as Union troops fled to Cemetery Hill. Soon outflanked on both right and left, many fought to the death until they were sure the bulk of the Union was safe.

With 1,200 casualties, out of a total of 1,800 who entered the battle, the Iron Brigade's losses had indeed been grievous; but by holding off the Confederates on the west of the town they enabled the rest of the Union army to place themselves in strong defensive positions south of the town on Culps Hill, Cemetery Hill and Cemetery Ridge, from where the Confederate army was unable to dislodge them. Thus the gallant Iron Brigade played an important role in deciding the outcome of the battle, and in the final triumph of the Union.

Gettysburg was sadly the Iron Brigade's last official stand. After the battle and the huge losses they sustained, men from many states of the Union were brought in to replace them. Although they played a valuable part under General Grant in the final advance on Richmond, this new unit could no longer be reckoned as the old Iron Brigade of westerners.

With the war ending once and for all on 9 April 1865, fewer than 200 of the original 1,026 men returned to Wisconsin. In Colonel William F. Fox's *Regimental Losses in the American Civil War*, he reported that the 2nd Wisconsin lost the highest percentage of men killed in battle of any regiment in the Union army in proportion to the number enlisted. But their efforts did not go unnoticed.

'Of all the brave troops who have gone from our state,' reported the Detroit *Free Press*, 'few, if any, regiments can point to a more brilliant record, to more heroic endurance, to greater sacrifices for the perpetuation of the priceless legacy of civil liberty and a wise and good government.' A correspondent for the *Milwaukee Sentinel* echoed these thoughts:

> Wisconsin, Michigan and Indiana can say with truth that they have furnished the bravest soldiers of the war and they have had their shoulders to the wheel ever since the rebellion broke out. Their soldiers have never faltered ... [and] they were confident that Right would be vindicated – and the result proved they were not wrong.

Such bravery on a battlefield is an enormous asset. When coupled with brilliant tactics, it can make a unit nigh-on unstoppable. And such a combination would be needed in abundance when a group of men had to do battle with a fleet of tanks, or see their country lose the greatest war of their age ...

18

THE STORMTROOPERS

AD 1914

As war broke out in Europe on 28 July 1914 and spread from the Balkans to become a global conflict, fighting on several fronts soon descended into trench warfare. On the Western Front, in particular, men fought in appalling conditions; casualties were alarmingly high and progress agonisingly slow, as neither side seemed able to break the deadlock.

Relentless infantry bayonet charges were useless as thousands of men were quickly cut down by machine-gun fire without gaining an inch of ground. Alternatively, attempts to annihilate the enemy with a massive artillery barrage proved equally ineffective. The enemy would merely wait in their underground dugouts until the barrage lifted before emerging to mow down the attacking infantry. As was seen so tragically at Verdun, where 337,000 died, it seemed both sides had only one strategy: to fling more firepower at the enemy than they could fling back. Indeed, at the Battle of Arras my 21-year-old uncle John was also to die in such circumstances when leading a company of the 2nd Battalion of the Gordon Highlanders against the German trenches.

Despite the horrendous number of casualties, there seemed to be no way to break the stalemate. Each day that passed merely saw each side's defensive positions get stronger, with barbed-wire obstacles installed across no-man's land, as well felled trees with sharpened branches and pointed metal stakes. However, the British were about to make an almighty breakthrough and unleash one of the most fearsome weapons the battlefield had ever seen.

On 20 November 2017, over a 1,000 guns suddenly opened fire on the German trenches defending the town of Cambrai, followed by smoke and a creeping barrage of British forces. Despite this onslaught, such an attack was par for the course. But what came next was not. Under cover of this ferocious bombardment, 376 Mark IV fighting tanks suddenly lumbered across no-man's land, spearheading a surprise attack that crushed the belts of defensive barbed wire and smashed through the German lines. The tank's armour allowed it to shrug off bullets and shrapnel, while its weapons knocked out German machine-gun nests with impunity.

Most of the Germans had never seen anything like it before. Cambrai was the first battle in which tanks were used en masse and they were a terrifying sight. When the Germans set eyes on these killing machines storming towards them, they fled for their lives.

Within twenty-four hours, the British had driven the Germans back 5 miles – a feat unprecedented on the Western Front – as well as destroying two German infantry divisions. After three years of trench warfare, the British Army at last looked to have the Germans on the run. At this momentous news, church bells rang out in Britain for the first time since the outbreak of the war.

Germany now appeared to be staring defeat in the face. It had no tanks of its own and it seemed that this new machine would sweep all before it and make trench warfare obsolete. But, while the British had been developing the tank, the Germans had been working on trench-clearing tactics of their own. Now it was time for the acid test. Their *Sturmtruppen*, aka 'storm troops', were their last hope.

The origin of the storm troops can be traced to 2 March 1915, when the German war ministry ordered the 8th Army Corps to form an assault detachment (*Sturmabteilung*). The idea was to attack by stealth and penetrate weak points in the enemy line, without expending blood or time attacking enemy strongpoints. In theory, by the time the defender realised what was happening, his front-line forces would be surrounded and isolated, to be mopped up later by follow-up waves of regular German troops. Under the command of Major Calsow, this fledgling new unit subsequently assembled at the artillery range at Wahn, where it spent the next two months honing techniques for storming trenches.

Issued with lightweight Krupp guns, which were soon christened the 'assault cannon', as well as portable steel shields, they learnt how to clear barbed wire and other obstacles in no-man's land. However, this training was never put into effect. Rather than lead an assault they instead defended a section of the German trench line. Casualties were high, while the assault cannon was found to be unsuitable for use near the front line. Each time one was fired its pronounced muzzle would flash, making it easy for the enemy to determine its exact position. Major Calsow was subsequently relieved of his command, to be replaced by the 37-year-old Captain Rohr of the Guards Rifles Battalion.

Rohr was given a free hand insofar as the training of his unit was concerned. His only instruction was to train his unit 'according to the lessons that he had learnt during his front-line service' with the Guards Rifles Battalion. This was to have a profound effect on the future of the assault detachment. In the next few months, Rohr was to transform it into an elite infantry organisation.

One of the first things Rohr addressed was weaponry. Rather than be restricted to bayonets or cannons, Rohr believed an assault battalion needed an array of weapons suitable for any given situation. As such, he was provided with a machine-gun platoon (two model 1908 Maxim machine guns), a trench mortar platoon (four light mortars) and a flamethrower platoon of six small flamethrowers.

The subsequent methods taught at Rohr's courses also consti-
tuted a radical departure from the initial trench-clearing tactics
of 1914. Squads were now treated like tactical entities in their
own right, moving as individual units towards their predesignated
objectives. In their movement across no-man's land, no attempt
was to be made for the squads to maintain any sort of connection
with each other. Neither was there to be any sort of predetermined
formation within the squad. The men would move so as best to
take advantage of the terrain.

While this approach might seem haphazard, meticulous preparations went into each assault. To ensure detailed knowledge of the
battlefield and its many obstacles, the men were provided with
large-scale maps, while a full-scale model of the enemy position,
complete with trenches and barbed wire, was also built in the
training area behind the lines for dress rehearsals, some of which
included the use of live ammunition.

To ensure stealth, equipment was also modified. The heavy hob-
nailed leather jackboots, long associated with German infantrymen,
were replaced by lace-up half-boots and puttees of the kind used by
Austrian mountain troops. To facilitate crawling, the field uniform
was reinforced with leather patches on the knees and elbows. The
leather belt and shoulder harness that had supported rifle ammunition pouches were discarded in favour of a pair of over-the-shoulder
bags to carry hand grenades, the stormtroopers' weapon of choice.

However, that's not to say that the rifle wasn't still a key part
of their armoury. Initially, Rohr had adopted the Russian 76.2mm
field guns for his troops. A large number of these had been captured in Poland and the Ukraine in 1915, and were then modified
to optimise them for mobility. As such, barrels were cut down and
long-range sights, superfluous on a weapon expected to hit enemy
targets less than 1,000 metres away, were removed. In time, these
were, however, replaced with a Mauser carbine, which was lighter
and easier to handle.

For close-quarters fighting in the enemy trenches, the stormtroopers used an artillery model Luger, which boasted a much-increased 32-round magazine. As Erwin Rommel observed, 'In a man-to-man fight, the winner is he who has one more round in his magazine.' Eventually, the storm troop battalions would also receive the world's first effective sub-machine gun, the MP18, which introduced most of the features that were to make it the key close-quarters weapon of the Second World War. However, these were to arrive in 1918, after most of the fighting had been done.

Despite the many guns at their disposal, it was the flamethrower that proved to be the stormtroopers' most terrifying weapon of all. Crossing no-man's land, one man would carry the fuel tank on his back, while a second man aimed the tube at the enemy, spitting 40-metre-long streams of burning oil at them. First tested against the French at Verdun, the terror inspired by the jets of liquid flame enabled the German assault troops to capture their objectives with relative ease. Unsurprisingly, no man was prepared to remain in a trench with blazing fuel oil cascading over the parapet.

To defend his stormtroopers, Rohr experimented with various kinds of body armour. The portable steel shields that were already in the inventory of the assault detachment were initially utilised, as well as steel breastplates reminiscent of those of the late Middle Ages. However, Rohr soon found that they did not allow his stormtroopers to move fast enough. As such, the only piece of armour that Rohr adopted for all operations was the 'coal scuttle' helmet, which was to become the trademark of the German soldier of both world wars.

In early October 1915, the storm troop was put to the test in an assault on a French position in the Vosges mountains, known as the Schratzmannle. After a detailed rehearsal, flamethrowers led the assault, followed by stormtroopers clearing trenches with hand grenades. They were then supported by German trench mortars and artillery to silence the French guns. This combination proved to be irresistible.

However, it wasn't until 10 January 1916 that the storm troops would be used as a complete unit for the first time, with an assault on the Hartsmannsweilerkopf, a ridge in the Vosges mountains. The attack had all the same elements as the assault on Schratzmannle – a detailed rehearsal, individual squad movement and the close co-ordination of flamethrowers, machine guns, infantry guns and trench mortars. It also achieved similar results, as French positions were cleared with only light casualties. Later, in the Italian mountains of Caporetto, stormtroopers enabled a German–Austrian offensive to capture 265,000 Italians and nearly knocked Italy out of the war in the process.

It was now abundantly clear that storm troops could be the key to victory in trench warfare. As such, by the beginning of December 1916 the 1st, 2nd and 5th German armies each had an assault battalion, and the other fourteen German armies were to establish one during the course of the month.

However, the British tank attack at Cambrai in November 1917 had the German command rocked. Its only means of response was through the stormtroopers, but with time of the essence neither the storm troops nor the ordinary infantry that followed them had the opportunity to rehearse the attack. Rapid improvisation, and instinctive execution of battle drills, would have to suffice. Even so, was a unit of men really going to be of much use against the might of the tank? Its armour appeared impregnable and there appeared to be no way into the hold. But, after studying them relentlessly, the Germans soon had a breakthrough. They realised that, if they tied hand grenades together and targeted the tank's tracks, they might just be able to stop them dead.

Commencing their counterattack just ten days later, the stormtroopers raced across no-man's land launching a hurricane barrage of explosives and gas shells. Cloaked by smoke, gas and fog, they infiltrated the British strongpoints and cut off the dazed defenders, quickly overrunning the first trench system and then pushing forward, leaving the waves of ordinary infantry following behind to finish them off.

But then the British hit back. Taking up defensive positions in each village, wood or old communications trench, they were determined to make the Germans pay for every mile. Meanwhile, their tanks roared onto the battlefield, intent on forcing them back.

This time, the stormtroopers were ready for them. Tying their grenades together, the nimble troops raced either side of the tanks and placed their explosives on their tracks. The subsequent explosion saw the tanks judder to a halt and become stuck in the mud. Their guns were also silenced soon after, as stormtroopers stuffed grenades inside to blow them up. Following this, they dragged the drivers and gun operators out of the hatch and shot them dead.

Over the next forty-eight hours, the stormtroopers not only managed to recover all of the territory that had been lost to the British tanks, but they also took a considerable piece of ground that the Allies had held since the autumn of 1914. These gains were monumental, especially after what the British had achieved with their tanks. Just days after facing total defeat, the Germans now believed the stormtroopers might actually help them win the war. At this, the German High Command looked to further improve its tactics with a series of large-scale training exercises, led by General Oskar von Hutier, now commanding the 8th Army.

This refined approach involved the following procedure:

1. A short artillery bombardment, employing heavy shells mixed with numerous poison-gas projectiles, to neutralise the enemy front lines, and not try to destroy them.
2. Under a creeping barrage, the stormtroopers would then move

forward, in dispersed order, avoiding combat whenever possible, infiltrate the Allied defences at previously identified weak points, and destroy or capture enemy headquarters and artillery strongpoints.

3. Next, infantry battalions with extra-light machine guns, mortars and flamethrowers would attack on narrow fronts against any Allied strongpoints the shock troops missed. Mortars and field guns would be in place to fire as needed to accelerate the breakthrough.

In the last stage of the assault, regular infantry would mop up any remaining Allied resistance.

By the spring of 1918, with their tactics refined and hope restored, the Germans prepared to use their stormtrooper units for what they hoped would be their final breakthrough, ironically called the 'Peace Offensive'.

Following von Hutier's training, the operation began at 4.40 a.m. on 21 March 1918, with a preparatory bombardment of shells and gas on British positions, followed by a salvo from the light field batteries. From here, the stormtroopers went into action, with their goal to penetrate the 8,000 metres or so that stood between no-man's land and the British field artillery emplacements. They only had twelve hours of daylight to accomplish this, but they were aided by the thick fog and gas that had lain over most of the battlefield for most of the morning. As such, the British machine-gun teams could not see the grey-clad attackers until a few seconds before they were overrun. Like ghosts they moved through the mist at speed, taking down anyone who stood in their way, whether it be via gun or knife. They knew from their training where the British were meant to be stationed and relentlessly picked them off.

However, while the stormtrooper units again achieved startling successes in the days that followed, it was not enough to sustain victory. If anything, the tactics had been too successful. The German infantry was unable to keep up with them, leaving them isolated, while the stormtroopers became so exhausted that fatigue severely affected them at a time when they needed all their stealth and cunning. A tired man could be put in a dense formation and pushed into battle – that was one of the virtues of close-order tactics. The stormtrooper, on the other hand, had to provide his own motive force. By the third day of the attack, that motive force was wearing thin and troops were observed not only stopping to loot the British supply depots that were falling into German hands, but also getting drunk on captured rum.

By the ninth day of the attack, the will of individual German soldiers to push forward collapsed, and the British were finally able to reconstitute a line and start building a new defensive system. For six more days, the Germans tried, without success, to break through this new line, before the operation came to an end on 5 April 1918.

The gains that the Germans made during the Peace Offensive were impressive by the standards of the Western Front – 90,000 prisoners and 1,300 guns were taken, 212,000 enemy soldiers were killed or seriously wounded, and an entire British army (the Fifth) was destroyed. However, the cost had still been high. Casualties totalled 239,800 officers and men, with some divisions reduced to half-strength.

The failure of the Peace Offensive marked the beginning of a long, slow retreat, which lasted until November 1918 and heralded the end of the First World War. Germany lost the war but its stormtroopers had totally revolutionised military tactics and proven that man could still be effective against machine. Indeed, this was a lesson that many in the west continue to ignore, as seen by the disastrous forays into Vietnam and Afghanistan, where clever guerrilla tactics proved more than a match for top-class technology. Yet, when conflict was again to darken the skies of Europe just twenty-one years later, it would initially be a battle of two machines that would ensure the fate of the war for the years to come ...

19

THE RAF AND THE
BATTLE OF BRITAIN

AD 1940

Although Europe had been at war since Hitler's invasion of Poland
a year before, London had remained relatively untouched by the
conflict. While the Nazis had rampaged their way through the
continent, Britain had so far stood firm, even in the face of Hitler's
stated intention to invade the British Isles in an operation code-
named 'Sealion'.

And on this day the war could not seem further away. As the last
of the summer sun shone brightly over England's capital city, it shim-
mered off the Thames and illuminated the great sites of Buckingham
Palace, St Paul's Cathedral and Tower Bridge. Revellers soaked up
the rays in Hyde Park, while the West End theatres continued to do
brisk business. The novelist, William Sansom, recalled it as 'one of
the fairest days of the century, a day of clear warm air and high blue
skies'. But the peace was about to be shattered. Almost 300 years
earlier, the Dutch had raided the city by boat and now the Germans
looked to do so from the sky. It was the start of the Blitz.

Suddenly, air-raid sirens screamed out, sending London's citizens clamouring for shelter. Soon the shadows of hundreds of

German bombers and fighters covered the ground, the distinctive roar of their engines filling the air. Opening their bellies, they proceeded to empty their steel shells of their loads. Seconds later, hundreds of bombs hurtled to earth and slammed into the factories on the Thames. Within moments, London was ablaze. American newspaperman Ben Robertson, who watched the attack, wrote that London, which 'had taken thirty generations of men a thousand years to build', was now in the process of being destroyed. It was the city's worst fire since the Great Fire of 1666.

With such devastating consequences, this attack only marked the start of Hitler's quest to cripple Britain and commence an invasion that would see him rule all of Europe. Unless Britain could defend its skies then it would soon be in German hands. But, against the might of the Luftwaffe, many thought Britain didn't stand a chance.

While the Germans had been banned from operating a military air force following the First World War, the subsequent rise of the commercial airline Lufthansa had allowed Germany to continue to train pilots and be at the forefront of aviation technology. Moreover, they had also defied the Treaty of Versailles by secretly training pilots in glider clubs or in the Soviet Union. The rise of Hitler in the 1930s saw him ignore the prohibition altogether and proceed to rapidly expand the German air force.

Hermann Goering, a First World War fighter pilot, was appointed aviation minister, while Erhard Milch was appointed State Secretary of the Reich Ministry of Aviation. Under their command, Germany had a fleet of Dornier bombers and Me109 fighter planes, which were the envy of the world. Its pilots were not only highly trained, with candidates tested unrelentingly for months on end, but many had also picked up crucial experience in 1937 during the Spanish Civil War. On top of all this, by the summer of 1940, the Luftwaffe outnumbered the British fighter defenders by almost five to one.

Unsurprisingly, Hitler and Goering looked towards an invasion of Britain with increasing confidence. In the face of this threat, Joseph Kennedy, the American ambassador to London, evacuated his children back to the United States, while General Weygand, the defeated French commander-in-chief, dismissively commented, 'In three weeks, England will have her neck wrung like a chicken.' But Winston Churchill did not share this pessimism.

In a speech to the House of Commons on 18 June 1940, just days before France signed an armistice, Churchill famously roared, 'We shall fight on the beaches, we shall fight on the landing grounds, we shall fight in the fields and in the streets, we shall fight in the hills; we shall never surrender.' But, for now, it was in the air that Britain needed to fight, with Churchill optimistic that they could actually prevail:

> The great question is: can we break Hitler's air weapon? Now, of course, it is a very great pity that we have not got an Air Force at least equal to that of the most powerful enemy within striking distance of these shores. But we have a very powerful Air Force which has proved itself far superior in quality, both in men and in many types of machine, to what we have met so far in the numerous and fierce air battles which have been fought with the Germans ... I look forward confidently to the exploits of our fighter pilots – these splendid men, this brilliant youth – who will have the glory of saving their native land, their island home, and all they love, from the most deadly of attacks.

Britain had first deployed a military air force, known as the Royal Flying Corps, at the start of the First World War. With aviation technology still in its infancy, the corps was initially a small force of just sixty aircraft, but by the end of the war it boasted a fleet of over 23,000 new machines, as well as a new name – the Royal Air Force.

Over the coming years, Britain seemed to take the furthering of its air force seriously, with the world's first military air academy opened at Cranwell in 1920, complemented two years later by the creation of an RAF Staff College at Andover. In 1923, an ambitious if not realistic target was even set to establish fifty-two squadrons for the purposes of home defence.

However, this rapid progress soon ground to a halt. With the memories of war slipping away, by the end of the 1920s the RAF still only consisted of twenty-five home-based regular squadrons, supplemented by eleven auxiliary and reserve units.

Desperate to jump-start the process, in 1934, the government voted for a five-year general expansion plan for the RAF. In 1936, the service was subsequently divided into four key commands: Bomber, Fighter, Coastal and Training. Hugh Dowding was appointed commanding officer of Fighter Command and would have direct control over the fighter force, anti-aircraft and balloon commands, as well as the Royal Observer Corps.

Dowding had been an aerial prodigy almost from the moment aircraft were invented. Having attended the Royal Military Academy in Woolwich, after which he was commissioned as an officer in the Royal Garrison Artillery, he had soon developed a keen interest in flying. Going on to join the Royal Flying Corps, he received his wings after just three months' tuition and was soon thrown into action in the First World War. With regular combat in France, his rise within the fledgling RFC was meteoric. After commanding No. 16 Squadron, he was posted to Training Command, where, in the post-war RAF, he gained experience in departments of training, supply, research and development, before his promotion to the rank of air marshal in 1933.

Yet, despite all of this, Dowding's promotion to Fighter Command was not universally welcomed. He was accused of not being a 'people person', and it was this alleged inability to charm that subsequently saw him overlooked as candidate for the role of chief of the air staff, the most senior position in the RAF. Meanwhile, his junior in both age and experience, Cyril Newall, was promoted ahead of him. Dowding was soon seen as yesterday's man and he was earmarked for retirement in June 1939. However, he was to be reprieved. Due to the looming threat of war, Newall asked him to stay on for an additional year. As such, when Germany looked to invade during the summer of 1940, Britain's defences were in the hands of someone who had been all but consigned to the scrapheap just a few months before.

However, notwithstanding what others might have thought of his personality, Dowding had ensured that he, and Fighter Command, were ready for the task. In 1936, he had already begun to implement and refine a defensive screen, which would later be known as the 'Dowding system' and would prove vital to Britain's defences. This system saw Fighter Command divided into four operational groups, each consigned to different sectors of the UK. Each group had its own headquarters that would send out orders to individual sector stations based at key airfields. Upon these orders, the planes would be sent into the sky to do battle. The nerve centre for Fighter Command was situated at Bentley Priory, on the outskirts of London, where the initial fighter pilot orders were in the hands of the Women's Auxiliary Air Force (WAAF).

With the RAF suffering from an acute shortage of manpower at the outbreak of the war, women began to supplement, and then take over, roles that would normally have been denied to them, including as radar operators and plotters. From radar screens at Bentley Priory, the WAAF could see any incoming enemy aircraft and inform the relevant group of their positions. This system relied heavily on RAF wireless interception stations, which would take advantage of the poor radio discipline of German pilots and help to confirm the range and destination of enemy raiders.

However, while all of these elements contributed significantly to Britain's defences, they would be nothing without the planes and pilots who would actually do battle. And it would be the fighter planes under Dowding, rather than bombers, that would be crucial to Britain's success – in particular, the Hurricane and the Spitfire.

A mix of ancient and modern technology, the Hurricane had a wooden framework that was covered in fabric, while its Rolls-Royce PV twelve-piston engine was capable of speeds in excess of 330mph. This combination of old and new was ingenious. Economical, easy to produce and maintain, the Hurricane would prove to be the workhorse of Fighter Command. Even its somewhat old-fashioned elements would prove useful: enemy cannon shells, so quick to destroy all-metal constructions, were less deadly against the Hurricane as they simply passed through the fabric-and-wooden frame instead of exploding on impact. It was also an awesome killing machine. With its eight guns, fixed as two groups of four, it was perfectly suited to tackling the waves of German bombers that would soon be descending on Britain.

Nevertheless, while the Hurricane would outnumber the Spitfire by two to one throughout the Battle of Britain, it was the Spitfire that would become the iconic symbol of British resistance.

Its inventor, Reginald J. Mitchell, an acclaimed aircraft engineer, had started drawing up the designs in 1933. Following a trip to Germany, where he witnessed the rise of Adolf Hitler, he could tell at that early stage that another war was inevitable. On his return to Britain, he immediately went to work on designing an aircraft that could be useful in a war situation. As a result, Mitchell aimed to create a well-balanced, high-performance fighter aircraft, capable of fully exploiting the power of the Merlin engine, while also being relatively easy to fly.

By 5 March 1936, Mitchell had a prototype. With the Spitfire's distinctive elliptical wing giving the aircraft a higher top speed than several contemporary fighters, including the Hurricane, the RAF was blown away. As John Nichol writes in his fantastic

Spitfire: A Very British Love Story, 'The RAF had woken up to the demands of modern warfare. They needed a single-engine, single-seater fighter armed with eight machine guns that could produce the greatest destructive power possible in one quick attack.'

This saw the Air Ministry place an order for 310 of the new fighters. But, sadly, Mitchell was never to see his invention in combat, dying of cancer at the age of forty-two in 1937. His colleague Joseph Smith subsequently took over as chief designer, pushing it through a series of developments, which soon saw it find favour with its pilots.

George Unwin, a Spitfire pilot, enthusiastically told Nichol:

It was a super aircraft, it was absolutely. It was so sensitive on the controls. There was no heaving, or pulling and pushing and kicking. You just breathed on it and when you wanted, if you wanted to turn, you just moved your hands slowly and she went ... She really was the perfect flying machine. I've never flown anything sweeter. I've flown jets right up to the Venom, but nothing, nothing, like her. Nothing like a Spitfire.

Yet, with Britain already facing a shortage of aircraft, the Spitfire and the Hurricane did not roll off the production line as quickly as had been anticipated. Production was especially hampered by the difficulty of assembling an aircraft as precision-engineered as the Spitfire, which took an astonishing 13,000 hours to build. With time of the essence and resources running low, Lord Beaverbrook, a Fleet Street baron and also Minister of Aircraft Production, launched a national appeal on 10 July 1940. Placing an advert in all the national newspapers, he urged Britons to donate whatever they could that was made of metal, such as pots and pans, so that these items could be used in aircraft production. Throughout the country, every echelon of society subsequently stripped their kitchens bare. Beaverbrook also ran a Spitfire fund whereby Britons could contribute to the £20,000 cost to build just one aircraft. By the middle of August, over £3 million had been raised. A further 1,500 Spitfires would be airborne directly as a result of these appeals, each proudly bearing the name of their contributing fund on the fuselage.

But, even with sufficient aircraft, there was still a chronic shortage of skilled British pilots. Fortunately, Fighter Command could look to the Commonwealth for additional manpower. Soon men flocked to the cause from New Zealand, Australia, South Africa, Rhodesia, India and even Jamaica, as William Strachan had done so, becoming the only black man serving in the RAF. Pilots also came from America and Canada, as well as those countries that had already fallen to Germany, such as Poland, which contributed 146 pilots. Indeed, the commander of No. 11 Group, Dowding's deputy, Keith Park, was a New Zealander, while Quentin Brand, the commander of No. 10 Group, was South African.

Training such candidates was not an easy task. With the RAF quickly needing an influx of pilots, it meant that corners were inevitably cut in order to get them into the air. In 1939 alone, the RAF suffered 219 training fatalities, while many a pilot would be thrown into the front line having had barely a dozen hours' training in a modern fighter. In comparison, Luftwaffe pilots had an average of thirteen months of training and over 200 flying hours.

But Britain had no time to waste.

In July 1940, German aircraft moved on from targeting British supply ships in the English Channel and now aimed to hit its ports and harbours. With Britain's undermanned and undertrained pilots taking to the air to engage, vicious dogfights saw the loss of planes and pilots quickly mount up on both sides. In Nichol's *Spitfire*, pilot Hugh Dundas recalls his panic after his fuselage had been hit by cannon shells:

Smoke filled the cockpit, thick and hot, and I could see neither the sky above nor the Channel coast 12,000ft below.

Centrifugal force pressed me against the side of the cockpit and I knew my aircraft was spinning. Panic and terror consumed me and I thought, 'Christ, this is the end.' Then I thought, 'Get out you bloody fool; open the hood and get out.'

Desperately pulling himself out of the cockpit, Dundas was able to deploy his parachute and escape just in time, watching his Spitfire hit the ground below and explode as he sailed to earth. Many were not so lucky. Some were unable to escape their cockpits and burnt to death. Others who did manage to escape then found themselves entangled in their parachutes in the Channel and drowned. The glamour of aviation and war was soon swept away. Yet the British hoped that, if they could just delay the Germans until the onset of winter, then this would make it very difficult for them to continue with their daylight raids.

The British also had one major advantage when it came to frustrating any German attack. The German Me109 fighter planes escorting the bombers were operating at the extent of their fuel range. They therefore had only a limited amount of time to engage before being forced to turn for home. If the RAF could frustrate a German raid for long enough, then it would have to abort its mission.

As the pilots gained more battle experience, they also came to realise the most effective methods to take down a German plane, as listed in Sailor Malan's 'Ten Rules of Air Fighting':

1. Wait until you see the whites of his eyes. Fire short bursts of one or two seconds, and only when your sights are definitely 'ON'.
2. While shooting, think of nothing else, brace the whole of your body, have both hands on the stick, concentrate on your ring sight.
3. Always keep a sharp lookout. 'Keep your finger out!'
4. Height gives you the initiative.
5. Always turn and face the attack.

6. Make your decisions promptly. It is better to act quickly even though your tactics are not the best.
7. Never fly straight and level for more than thirty seconds in the combat area.
8. When diving to attack the enemy, always leave a proportion of your formation to act as top guard.
9. INITIATIVE, AGGRESSION, AIR DISCIPLINE and TEAM-WORK are all words that MEAN something in air fighting.
10. Go in quickly – punch hard – get out!

With the Spitfires and Hurricanes continuing to keep the Luftwaffe at bay, Hitler now called for the destruction of the RAF, with its radar stations along the coast of southern England key targets. Although the Luftwaffe was able to hit those stations regularly, they were only out of action for a few hours at a time, with frantic repairs soon making them operational again.

In addition, the Germans also began to target British airfields. On 30 August, Biggin Hill suffered particularly devastating consequences, when thirty-nine ground staff were killed, with twenty-six wounded, while the 500kg bombs destroyed armouries, storerooms and staff accommodation. The next day the Germans struck again, this time hitting the operations room itself. For the first time since the Battle of Britain began, a vital component of the defensive system was shut down. However, by the following morning it was up and running again, after engineers worked through the night to fix it.

No matter what the Germans did it seemed nothing truly crippled the RAF or its systems. In the face of such resistance, and with autumn soon approaching, Hitler was growing increasingly frustrated. He knew that Operation Sealion could not commence unless air superiority had been achieved over the southeast of England and yet nothing seemed to be working. The Germans needed a quick knockout blow to shake the life out of the British.

As such, they decided to target London. This would force large numbers of British fighters into the air, which the Luftwaffe then hoped to destroy in one big hit. Moreover, with the citizens of London also feeling the full wrath of the Luftwaffe, Hitler believed the British government would finally be forced to the negotiating table.

On 7 September 1940, RAF radar screens lit up as wave after wave of enemy planes crossed the Channel. Nothing like this had ever been seen before. With the WAAF sending out urgent orders from Bentley Priory, propellers were soon whirring on the airfields of every squadron within a 70-mile radius of the capital. But, before the British squadrons could attack, the first bombs were already raining down on the docks and factories along the Thames. As the air-raid sirens screamed their warning, people fled from the London sunshine to find shelter. For the first time since the start of the war, London was ablaze; 448 lives were lost in a single day. With the daylight bombings of London, everyone's worst fears of the Nazi war machine had been realized.

However, there was no way the Luftwaffe could maintain this kind of momentum on a daily basis. Over the next couple of days, only limited raids were launched, and the weather had finally begun to turn, grounding aircraft on at least one occasion. The British pilots were also now fighting with renewed vigour to protect their friends, their families, their homes and their country from obliteration. They would not be found wanting.

On 9 September, with the weather improving, the German raiders gathered in substantial numbers for another attack. But this time the RAF was ready for them. As the Luftwaffe approached in a pincer movement, with two separate raiding parties looming towards London via Dover and Beachy Head, the call quickly went from the WAAF to No. 11 Group. Taking to the skies, the stammer of the Merlin engines turned into a roar as they turned towards London and were positioned well in advance of the capital.

The first German raiding party was harassed by Spitfires, with the British planes swarming around them like a plague of wasps, eventually forcing the bombers to offload on Canterbury rather than London. The second formation was also frustrated. So much so that, in a desperate retreat, it also showered the surrounding towns and countryside with its arsenal. Despite high RAF losses – a total of seventeen planes and six pilots – the Luftwaffe had failed to strike at their intended target. Indeed, German losses were even higher – twenty-four aircraft and ten pilots – and with nothing to show for it.

Hitler was apoplectic. They could not withstand such losses indefinitely and, with winter not far off, there would not be many more opportunities to claim the skies of Britain. With this in mind, on 15 September Goering authorised one final attempt to destroy Fighter Command. For the Germans, it was now or never.

That morning, Churchill had chosen to visit RAF Uxbridge and so was present as the drama began to unfold. The first raid was spotted shortly after 11 a.m., with squadrons quickly scrambled from all over the southeast. It was clear that the Germans meant business. But the British squadrons were not deployed en masse. Two were situated well forward, hovering over Canterbury, another four patrolled over Biggin Hill, while a further two were held slightly in the rear as support.

As the German Dornier bombers, and their fighter escorts, crossed over into Britain, the British squadrons began to swarm around them like a pack of hunting dogs. Slowly the bombers' fighter escorts were stripped away as they became embroiled in fighting off the attacking Spitfires and Hurricanes. In response, the bombers bunched tightly together, offering each other protection, and continued on their flight path towards London. But before them was an almighty horde of British Hurricanes and Spitfires crowding the skies. Ducking and diving, the British soon shot down six bombers, with one abandoned by its crew, which plummeted towards the heart of London and crashed into Victoria Station. When one Dornier bomber looked to be heading for Buckingham Palace, it was intercepted by pilot Ray Holmes, whose encounter was covered by the *Sunday Express*:

He was hurtling straight for the Dornier. In a moment, he must break away. But the German pilot had not deviated an inch from his course. There was only one way to stop him now. Hit him for six. In the heat of the battle, with his own machine crippled and in a desperate bid to smash the invader before it broke through to his target, he shunned the instinct to turn away. How flimsy the Dornier tail-plane looked as it filled his windscreen. The tough little Hurricane would shatter it like balsa wood. As he aimed his port wing at the nearside fin of the Dornier's twin tail, he was sweating. He felt only the slightest jar as the wing of the Hurricane sliced through. Incredibly, he was getting away with it. The Hurricane was turning slightly to the left and diving a little. Suddenly, the dive turned vertical, Holmes was heading to the ground at 500mph. After a struggle Holmes managed to bail out. The scene of the Dornier, Hurricane and parachute coming to earth was watched by hundreds of grateful Londoners. The Dornier that seemed intent on bombing Buckingham Palace had been brought to earth. It was one of the many acts of heroism seen during the battle.

However, while the British looked to have succeeded in yet again frustrating the Germans, this was just a preliminary raid designed to tire the RAF before the Luftwaffe launched its main assault. As the bombers in the preliminary raid turned back, the 254 Hurricanes and Spitfires that had been deployed also returned to their bases to rearm and refuel. Just as they were doing so, the German main thrust was forming over Calais, consisting of 114 bombers and an incredible 360 fighters.

At Uxbridge, the commander of No. 11 Group, Keith Park, realised that the Germans were throwing everything they had at this. With Churchill watching on, cigar clamped between his lips, Park quickly issued orders to lull the attacking raiders into a false sense of security. They would hit the Germans with small attacks and then, just as their fighter escorts were beginning to run low on fuel, the British would launch their main assault, and hopefully cripple the Luftwaffe.

Shortly after 2 p.m., the first British defenders made contact. As Park instructed, there were only twenty-seven planes in the sky, but they still managed to strike down fourteen German aircraft, for only one loss. However, the majority of the vast German formation continued to bear down on London. Park now increased the number of attacks as he scrambled every squadron within No. 11 Group. But it was no use. The German bomber escorts swatted them away, with the bombers ploughing on in three columns towards their targets: the Royal Victoria and West India Docks.

However, between them and the docks were now the bulk of Park's defences – 185 fighters. They soon found themselves under attack and their fighter escorts became scattered as they twisted and turned, trying to escape the British fighters. Before long, their petrol lights had flicked on, just as Park had wanted, and now the main thrust of the British defenders dived into the attack. The German bombers desperately scanned the ground below for their designated targets but thickening cloud obscured most of the city directly beneath them. With just a few fighters to protect them, and unable to find their targets, the bombers were forced to turn for home, scattering their bombs over Dartford, Bromley and West Ham as they fled from their attackers.

As the weather worsened, and no further big raids appeared on the radar screens, the pilots of Fighter Command could finally relax. They had seen off one of the biggest air assaults that had ever been launched. But it had come at a cost. During the course of the day, the RAF had lost twenty-nine aircraft and twelve pilots had been killed. In contrast, the Luftwaffe had lost fifty-six aircraft and 136 men were either dead or POWs. There was a sense of jubilation in the air. The British pilots knew full well that despite the odds being against them they had countered everything that the Germans had thrown at them.

Across the Channel, the mood was despondent. For three months now, the German crews had been deluded by promises that Fighter Command was on its last legs, that the British were even down to their last fifty Spitfires. Yet again and again the defenders had pressed home their attack, in increasing numbers, and with a greater determination than before.

Time was now running out for Goering. It was already mid-September, and the glorious sunshine that had prevailed for most of the summer months could not last. With the weather turning, there would be fewer opportunities for the Luftwaffe to exploit. Hitler's patience was also wearing thin. His thoughts were now drawn eastward towards Russia. Goering's failure to achieve air superiority in time for an amphibious assault simply confirmed the inevitable. The launch date had already been moved several times; now, in the light of these substantial German losses, Sealion was postponed indefinitely on 17 September 1940.

The official end date for the Battle of Britain was 31 October 1940, as stipulated in the Air Ministry pamphlet on the subject published in 1941. In reality, war is not so clearcut and there was significant overlap. The German planes continued to wreak havoc, as night-time raids over London continued until May 1941, in the hope of forcing Churchill to surrender. Yet the British people did not abandon their homes, flee from their factories, or beg their government to negotiate a peace deal. They stayed strong and saw off the German attacks once again. The Blitz was to be a victory for the British spirit.

For Dowding, the battle had been won and now his war was over, officially retiring on 25 November 1940. Keith Park, his loyal and brilliant deputy, was also moved aside to make way for a new order at Fighter Command, with his rival, Trafford Leigh-Mallory, replacing him as commander of No. 11 Group.

That Germany failed to successfully invade Britain was one of the costliest strategic errors of the war. The maintenance of British sovereignty ensured that the island would later become a giant floating base for the Allied armies before the invasion of Europe four years later. Indeed, it is extremely doubtful whether the United States would have been able to wage a successful campaign in western Europe when separated by 3,000 miles of ocean.

Despite the many doubters, Dowding, and his brilliant defensive system, had emphatically proven them wrong. But the war was still not won. A series of key battles would be vital in banishing the scourge of the Nazis. One of which was to be known as the 'Greatest Raid of All Time', and it was another elite British unit that would again face impossible odds . . .

20

THE COMMANDOS

Following the failure to rule the skies in the Battle of Britain, Germany changed tack. Knowing that Britain, as an island, was somewhat reliant on importing food and materials in order both to feed its people and continue to prosecute the war, it now tried to dominate the seas. While German U-boats became the scourge of British ships, their battleships and heavy cruisers represented a real problem. These great leviathans could only be countered by other capital ships, or aircraft, and the availability of these weapons in the vast wastes of the North Atlantic was limited.

A particular threat to the British was the *Bismarck*. Along with the *Prinz Eugen*, it roamed the North Atlantic like a bully. On facing the pride of the British fleet, HMS *Hood* and HMS *Prince of Wales*, it took just ten minutes of action before the *Hood* was destroyed, sinking to the bottom of the ocean along with its crew. However, with the *Prince of Wales* subsequently outnumbered by the two German ships, she managed to get off a few shots of her own. One of these was to prove crucial. Hitting the *Bismarck*, it allowed the *Prince of Wales* to escape while the German ship had to urgently seek repairs. Fortunately for the *Bismarck*, the French port of St Nazaire, the only dry dock on the Atlantic coast capable of handling and repairing the great German warship, was not far away. But the *Bismarck* was not to get that far. En route it was destroyed by British ships, seemingly marking a monumental victory, but it was to be fleeting.

In Germany, an even more powerful ship, the *Tirpitz*, was now nearing completion. At 251 metres in length, and able to reach a speed of 30 knots, she was bigger and more powerful than any vessel the Royal Navy could muster. The threat that the battleship posed, and the realisation of what it could do to Britain's vital supply lines, became almost an obsession for the prime minister, Winston Churchill. The mere existence of the *Tirpitz* meant that four British capital ships had to be held in readiness at all times, waiting for her should she exit into the deep waters of the Atlantic upon her completion. Churchill told his chiefs of staff that no other target was comparable to the destruction of the great German capital ship. He even went so far as to say that the whole strategy of the war turned on her mere existence. At this, both the Royal Navy and the Royal Air Force immediately went to the drawing board.

Some looked at whether they could bomb the ship at port, or hit its weak spots should it ever enter the Atlantic. Others in the Admiralty, however, looked for a craftier solution, one that would not just stop the *Tirpitz* in her tracks, but also any other battle cruiser the Germans should build. They had noted how the ailing *Bismarck* had turned towards the port of St Nazaire after being hit. This was, after all, the only port in the North Atlantic that could handle a ship of that size. If the port was to be taken out of action, it was unlikely the German navy would risk the *Tirpitz* in the Atlantic, knowing it was the only dry dock where it could seek repairs. This was a masterstroke. Now all means had to be found to render the dry dock unusable.

In the meantime, Lieutenant Colonel Dudley Clarke, a staff officer at the War Office, had already come up with a solution, even if he did not know it yet. In June 1940, Clarke had drawn up the outline of a guerrilla raiding force that he called 'commandos'. His idea was passed to the prime minister, who very much liked what he read, subsequently ordering:

> Enterprises must be prepared with specially trained troops of the hunter class who can develop a reign of terror down the enemy coast ... I look to the Chiefs of Staff to propose measures for a ceaseless offensive against the whole German-occupied coastline, leaving a trail of German corpses behind.

Indeed, it is no surprise that Churchill was enthusiastic about Clarke's idea, or his name for the new unit. During the Boer War, Churchill had been a newspaper correspondent and had observed how Boer 'Kommandos' had conducted a series of hit-and-run raids that tied down large numbers of British troops. By the time the Boers were finally persuaded to make peace, the British had been taught invaluable lessons in irregular warfare that they would never forget. Moreover, Churchill had also seen just how effective guerrilla raiding tactics were during the First World War thanks to the exploits of T. E. Lawrence in Arabia against the might of the German and Ottoman armies.

However, not everyone shared Churchill's enthusiasm for this new unit. To many high-ranking service officers, commandos were anathema. They considered the new force to be a misuse of good officers and men who would have done better to remain with their units. In the view of these officers, the war would be won by conventional means, not by sideshows.

Nonetheless, just forty-three days after Clarke had suggested a raiding force, Admiral Sir Roger Keyes was made head of Combined Operations, and became responsible for the direction of all raiding operations and their co-ordination with naval and air forces.

The troops who were to form the commandos initially came from two sources: volunteers from the units of Home Commands as well as those from Territorial Divisions. Many flocked to the cause, excited by the prospect of 'hazardous service', and also the pay, which was 13s.4d. (68p) for officers and 6s.8d. (33p) for other ranks. By the standards of the day, this was generous. Moreover, the commandos would live in billets, rather than barracks, for which they received a daily allowance to pay for their food and lodging. Best of all, life in billets meant that there were no barrack guards, cookhouse fatigues, or many of the petty and irksome chores of life in a barrack room.

With their training initially haphazard, and differing from unit to unit, a uniform set of standards was eventually laid down by the commanding officer, Lieutenant Colonel Charles Newman, which read as follows:

1. The object of Special Service is to have available a fully trained body of first-class soldiers ready for active offensive operations against an enemy in any part of the world.

2. Irregular warfare demands the highest standards of initiative, mental alertness and physical fitness, together with the maximum skill at arms. No Commando can feel confident of success unless all ranks are capable of thinking for themselves; of thinking quickly and of acting independently, and with sound tactical senses when faced by circumstances which may be entirely different to those which were anticipated.

3. Mentally. The offensive spirit must be the outlook of all ranks of a Commando at all times.

4. Physically. The highest state of physical fitness must at all times be maintained. All ranks are trained to cover at great speed any type of ground for distances of 5–7 miles in fighting order.

5. Cliff and mountain climbing and really difficult slopes climbed quickly form a part of Commando training.

6. A high degree of skill in all branches of unarmed combat will be attained.

7. Seamanship and Boatwork. All ranks must be skilled in all forms of boatwork and landing craft whether by day or by night, as a result of which training the sea comes to be regarded as a natural working ground for a Commando.

8. Night sense and night confidence are essential. All ranks will be highly trained in the use of the compass.

9. Map reading and route memorising form an important part of Commando training.

10. All ranks of a Commando will be trained in semaphore, Morse and the use of W/T (radio).

11. All ranks will have elementary knowledge of demolitions and sabotage. All ranks will be confident in the handling of all types of high explosive, Bangalore torpedoes, and be able to set up all types of booby traps.

12. A high standard of training will be maintained in all forms of street fighting, occupation of towns, putting towns into a state of defence and the overcoming of all types of obstacles – wire, rivers, high walls, etc.

13. All ranks in a Commando should be able to drive motorcycles, cars, lorries, tracked vehicles, trains and motorboats.

14. A high degree of efficiency in all forms of fieldcraft will be attained. Any man in a Commando must be able to forage for himself, cook and live under a bivouac for a considerable period.

15. All ranks are trained in first aid and will be capable of dealing with the dressing of gunshot wounds and the carrying of the wounded.

16. These are few among the standards that must be attained during service in a Commando. At all times, a high standard of discipline is essential, and the constant desire by all ranks to be fitter and better than anyone else.

17. The normal mode of living is that the Special Service soldier will live in a billet found by himself and fed by the billet for which he will receive 6s.8d. per day to pay all his expenses.

18.   Any falling short of the standards of training and behaviour on the part of the Special Service Soldier will render him liable to be returned to his unit.

As the threat of a German invasion receded, the tempo of training increased, with the emphasis on raiding. Much of this was carried out at the Irregular Warfare School in Scotland, which ran courses for 'special forces' in the mountains and lochs of the wild Highland country. Endurance marches, cliff climbing, swimming with full kit, killing with and without weapons, seamanship and boat work, night operations, map reading, fieldcraft, stalking and much else was taught, at a pace that only the very fittest survived.

However, the first commando operations in 1941 achieved only varying success, although a raid of the Norwegian Lofoten Islands in March did see a number of wheels for a German Enigma encoding machine seized, which enabled the British to decode much German radio traffic over the course of the war. Despite this, the raids were often small, and unsuccessful, other than keeping the German forces on their toes.

As such, in October 1941, Captain Lord Louis Mountbatten was installed as the new chief of combined operations, in the hope that he could inspire the commandos to greater feats. But raids in Europe and the Middle East continued to be viewed largely as failures.

Nonetheless, it was the commandos to whom everyone soon looked to target the dock of St Nazaire, as all initial options were deemed unsuitable. The RAF had already attempted to bomb St Nazaire without much success, due to precision bombing not being an exact science. Moreover, there was also a reluctance to engage in a large-scale bombing campaign since it might cause large casualties to the nearby French civilian population.

In January 1942, with all other options ruled out, Churchill turned to his misfiring commandos. While Lord Mountbatten was tasked with planning the operation, he realised it would be extremely difficult. St Nazaire was located 5 miles up the treacherous estuary of the River Loire and was only approachable from the sea by a single narrow channel, which, in 1942, was covered by several batteries of coastal defence guns. Going up the River Loire would therefore result in certain detection well before any landing could be made. However, Mountbatten soon had a moment of inspiration.

He found that, in late March, there would be exceptionally high spring tides that would allow a vessel of shallow draught to pass over the sandbanks and bars that dotted the estuary of the Loire, rather than approaching the docks along the well-protected shipping channel. It was an opportunity too good to miss. But how could a small unit carry enough explosives to totally destroy the dock? This was when Mountbatten now had a true masterstroke.

He realised that, if they could pack a ship with enough explosives and ram the dock gates, this would certainly destroy the dock. As such, HMS *Campbeltown* was loaded with 4 tons of explosives, encased in steel and hidden within her bows. But this was not a suicide mission: the charges were designed to explode hours after the collision, so as not to kill all the commandos on board.

To avoid detection, the ship would be disguised as a German destroyer. However, if it should come under attack it was reinforced with extra metal plating and armed with cannons. There was also concern that it might be too heavy for the shallow waters, so it was stripped of all non-essential items, to make it as light as possible.

Following the ramming of the gates, supporting ships would deposit teams of commandos onto land. From there, they would destroy key targets, and then board the waiting ships to return home. It was a high-risk plan, with little chance of success, but it was all Churchill had. As such, he gave it the go-ahead.

Lieutenant Colonel Newman was subsequently chosen to lead the raid. A building contractor by profession, Newman was a pre-war territorial officer from the Essex Regiment, whose thirty-eight years made him seem very old to his younger subordinates, most of whom were in their early twenties. However, his skills in leadership, and his ability to relate to his men, made him a popular and well-respected commanding officer. After being given command of 2 Commando, Newman spent the next year preparing it for the raid of a lifetime: Operation Chariot.

A crucial part of the raid would be played by the demolition teams, who would lay explosives at key parts of the dock. These teams were sent to Burntisland, on the shores of the Firth of Forth, to undertake a specialised course in the destruction of dockyards. Trained in the handling of modern explosives, the men were familiarised with the weight and shape of the charges they would have to place and taught how to identify the precise spots where the explosives would have to be positioned to gain maximum effect. At Southampton and Cardiff docks, they practised over and over, against the clock, in the dark, and often without key members of their group, who were suddenly deemed to be casualties.

Their targets at St Nazaire were:

1. The destruction of the two caissons of the Normandie Dock;
2. The demolition of the dockyard facilities supporting the dry dock, such as the winding sheds and the pump house;
3. The wrecking of all lock gates; and
4. The destruction of any shipping that was present, especially U-boats.

It was vital that these demolition teams were protected as they went about their job. Over 100 of Newman's best men were selected for intensive training in securing and holding crucial demolition positions, while keeping the enemy at bay.

The Royal Navy were also involved in the operation, having to actually place the commandos at St Nazaire and then get them back home safely. The man chosen to command the operation was 34-year-old Robert Ryder. Ryder was, however, a surprising choice. At the time, he was languishing in a desk job in a stately home in southern England, suffering the Admiralty's 'displeasure' at having lost his last ship. This therefore offered the rare opportunity for redemption.

On the afternoon of 26 March 1942, the explosive-packed HMS *Campbeltown*, disguised as a German destroyer, and its fleet of small boats left Falmouth and made their way into the Channel bound for France. It was a mission that could determine the fate of the war.

However, the journey across the Channel was fraught with danger. On the morning of 27 March, they crossed paths with a German U-boat that had risen to the surface. Avoiding conflict was vital to ensure that St Nazaire wasn't notified of the incoming commando invasion. Thankfully, the commanders of HMS *Campbeltown* were well versed in German signals. This bought them some time as the U-boat tried to work out if the flotilla was friend or foe. But, with any response seeming to take an eternity, the British boats became nervous. Opening fire, they caused the U-boat to dive and attempt to escape. As a game of cat and mouse ensued, the escorting destroyer, the *Tynecastle*, dropped a pattern of depth charges across the area but could not verify if the U-boat had been hit. The bombs had in fact missed, and the U-boat had escaped, but it reported that the flotilla looked to be heading for Gibraltar rather than St Nazaire. For now, at least, the route remained clear.

As the day wore on, the sky became overcast and the clouds descended to give some much-needed cover to the raiding force. It now made a steady 8 knots and pressed on towards the French coast. At 8 p.m., they began to close directly on the port.

To provide cover, as well as a distraction, at about midnight the RAF began to drop bombs on St Nazaire. As the bombers droned overhead, the boats of Operation Chariot passed through the dark waters towards the estuary of the River Loire. All was quiet, with no searchlights probing the sea.

Entering the wide mouth of the river at fifty minutes past midnight, the convoy passed the ghostly wreck of the liner *Lancastria*, sunk in 1940, the scene of the greatest loss of life in British maritime history. Yet something else soon caught their attention. Up ahead were German search towers. Nevertheless, the disguised HMS *Campbeltown*, and its convoy, sailed past them unopposed. All German eyes looked to be directed towards the RAF. The only issue the convoy seemed to face was the possibility of running aground in the shallow waters. But for now *Campbeltown*'s momentum carried her forward whenever she did touch the bottom.

However, the German command at St Nazaire soon grew suspicious of the RAF's movements. At this, they put out a call to all units to be on the lookout for any enemy boats. The game was soon up. All searchlights on both sides of the river immediately came to life and picked out the *Campbeltown*. Again, the Germans hesitated opening fire. The ship looked to be a German destroyer and they did not want to sink one of their own. When challenged about its intentions, the *Campbeltown*'s signalman flashed back that it was proceeding to St Nazaire as two of its craft had been damaged by enemy action. Still it was allowed to continue, and was now just six minutes from its target, entering the Loire proper. But the Germans were now certain that the convoy was hostile.

Soon every German battery and gun placement opened fire. In response, the *Campbeltown* and its convoy blasted back with all it had. They just had to survive a few more minutes to reach their target, but the barrage of gunfire was unrelenting. The blasts tore through commandos and crew, and the boiler and engine rooms took direct hits, bringing some ships to a standstill and making them sitting targets. At this, the casualties rapidly mounted and the impossible mission looked to be slipping away.

However, with the *Campbeltown* still thrusting forward under the relentless bombardment, a call suddenly went out, 'Stand by to ram!' Everybody on board now braced themselves for the shock of collision. Ahead, looming out of the darkness but lit with the flashes of guns and swept by searchlights, was the low dark strip of black steel that marked the entrance to the Normandie Dock. Suddenly, the 1,000-ton *Campbeltown* hit the anti-torpedo net that protected the lock. The warship's momentum saw it tear through the steel mesh and the destroyer leapt forward unchecked. Seconds later, with a grinding low groan, it hit the gates of the dry dock at 20 knots, lodging itself in such a position that the 4 tons of explosives in her forward compartments rested right up against the wall of the caisson. The charge could not have been better placed if it had been positioned there by hand. They just had to hope it would detonate as planned in a few hours' time.

But there was no time to think about that. Now the commandos on the *Campbeltown* and the other ships in the convoy had to somehow get onto the dock, all the while under an almighty barrage of gunfire. Some were already wounded on deck, but determined to press on, while others had no choice but to jump into the water and swim for it, as their boats were unable to land. Many were shot, others drowned, but soon small teams of commandos were on dry land, engaging with the enemy, dodging gunfire and racing to their target points to set off explosives and damage whatever they could.

Commando Corran Purdon told Sean Rayment in his book *Tales from the Special Forces Club*:

> As we emerged into the night, the whole of the sky seemed to be illuminated by tracer fire. The battle going on around us

was tremendous. The noise was deafening . . . There was blood all over the place but we hardly gave it a second thought. One chap, Johnny Proctor, whose leg had more or less been blown off, was cheering us on.

Such was the intensity of the fighting that it took on a surreal quality, as Hugh Arnold recalls:

It was the only time in my experience of the war that the reality approached the fiction that you see on television . . . The enemy had an enormous advantage. We couldn't really see anything. The searchlights blinded us. All we could do was to fire at the searchlights.

The pump house, and its propeller pumps that emptied the dry dock, was a particularly important target. However, the commando team responsible for this was in trouble. Its leader, Lieutenant Chant, had already been wounded in the legs. Despite the agonising pain and being under attack, Chant courageously led his team to the doors of the pump house, which they proceeded to blow open. The countless dry runs in Cardiff and Southampton soon proved invaluable as everything inside was just as they had planned. Under flickering lights, Chant and his men descended to the basement and set about placing the charges at strategic points on the pumps. With the explosives set, all that remained was to ignite the ninety-second fuse and run for it.

Despite Chant's wounded legs, he was the last to leave, ensuring the fuses were lit. Then, straining everything he had, he staggered up the stairs in the darkness, as the seconds ticked away. Just as the charges exploded, tearing the pump house apart with a roar that was heard throughout the dockyard, Chant threw himself out of the door. Bloodied, bruised and panting hard, he looked back with satisfaction at the scene of total devastation. Soon further explosions followed as other commando teams blew up the winding sheds and the dock gates, but things were growing increasingly precarious.

Out on the river, craft were sinking, boats were on fire and blazing pools of petrol and blood had spread across the water, with the cries of wounded men screaming for help reverberating over the sound of gunfire. The means of evacuation for the onshore commandos was also gradually being destroyed.

Yet still the mission continued, as Ryder gave the word for torpedoes to be fired at the outer lock gates of the submarine base. With a great roar, the motor torpedo boat ran at speed towards the closed dock and launched its missiles through the water, crashing them against the gates. With a reassuring clang, the torpedoes sank slowly to the bottom to await the moment when their time-delay fuses would activate.

With his part of the mission complete, it was now clear to Ryder that any successful evacuation plan for most of those commandos on land was going to be out of the question. He had no choice but to order the surviving boats to evacuate, leaving the commandos behind.

On land, Newman was now aware that they were on their own, and that most of the men who reported to him were injured. As Purdon told Rayment, upon reaching Newman, 'I said, "Sir, we have destroyed the northern winding house and we are ready to return to England." The colonel turned to me and said, "I'm afraid, old boy, the transport home has rather let us down. Have a look at the river."'

There were two options available to them: flight or surrender. Everyone agreed that surrender was out of the question. They would go down fighting trying to escape. Sadly, they were unable to take their wounded compatriots with them, having to leave them behind to face the prospect of a POW camp.

At this, Newman split up the surviving commandos into teams of twenty and told each of them it had to engineer its own way out of the docks, through the town and into the open country, with the aim of making for Spain and on to Gibraltar. It was a long shot but it was the only option left to them.

As the groups tried to make their way out of the docks, many were shot down, and others had to surrender, unable to move any further due to their wounds. While some crept into hiding places, hoping to move on once the fuss had died down, others somehow made it past the enemy and into the town, where German troops were now on patrol. Many of the men were apprehended but Newman and fifteen others, including Corran Purdon, sheltered in a cellar to wait for nightfall. But they were soon discovered, as were virtually all the others who were scattered through the town. However, against all odds, five commandos somehow evaded the Germans, made their way to Spain and then back to England. They were the lucky ones. Of the 611 men who had entered the Loire in the early hours of 28 March 1942, 169 were killed. But their losses were not in vain.

With the coming of daylight, the results of the raid could be clearly seen. Fires burnt everywhere, buildings were destroyed and ships were sunk in the harbour. On the river, the bodies of the dead floated downstream and were washed up all along the shores of the Loire. Any survivors were pulled from the river by German boats and sent to captivity. Meanwhile, the *Campbeltown* remained stuck fast on the outer caisson, her bows pointing sky-wards and stern settled in the mud.

As the hours passed, swarms of German sightseers climbed on board, along with naval troops and experts. Not checking the boat properly, they reasoned that its aim had merely been to ram the dock gates. It never occurred to them that the *Campbeltown* might be loaded with explosives.

By the middle of the morning, hundreds of curious people had arrived to take a look at the vessel. However, at 10.35 a.m., the pencil fuses, set by the now dead Lieutenant Tibbits, fired the 4 tons of depth charges in her bowels. The massive explosion that followed split the destroyer in two, while the caisson on which she lay collapsed into the dry dock.

While the explosion caused debris to cascade down across the town, on board the ship, hundreds of Germans were blown to oblivion, their body parts festooning the cranes and masts around the dockyard. Operation Chariot had achieved its prime objective, and then some.

With this explosion, a new wave of panic and paranoia overtook the Germans. They seemed to see enemies on every street corner and false reports that concealed British troops were pouring into German headquarters soon swept the town. Two days later, their paranoia went into overdrive when the delayed-action torpedoes blew up the outer lock gate to the submarine base. The Germans truly went out of their minds at this, with the British also now winning the psychological battle.

The Normandie Dock and its port installations were rendered completely useless to the Germans for the rest of the war. The *Tirpitz* never did venture out into the Atlantic and was finally destroyed by the RAF in a Norwegian fjord. The commandos had pulled off one of the most daring raids in military history. It was clear that unconventional warfare by elite soldiers could ensure monumental results, and this was something the Germans looked to replicate, albeit in the face of one of their commanders trying to sabotage their missions ...

21

HITLER'S
BRANDENBURGERS

AD 1942

It is quite ironic that the man put in charge of one of Germany's most controversial special forces, the 'Brandenburgers', despised Hitler and actually tried to sabotage the regime from within.

Fluent in six languages, Admiral Wilhelm Canaris had won the Iron Cross during the First World War for his intelligence work. With a stellar reputation, Canaris's work in post-war Osaka, overseeing the secret construction of the U-boat, saw him appointed as head of the *Abwehr*, one of six different intelligence services in Hitler's Third Reich, in 1935. However, the rise of Hitler made Canaris very uneasy. His subordinate, Erwin von Lahousen, would later say of him during testimony at Nuremberg:

He was a pure intellect, an interesting, highly individual and complicated personality, who hated violence as such and therefore hated and abominated war, Hitler, his system and particularly his methods.

As such, Canaris began to explore ways to undermine the nation's leadership in order to prevent another conflict. His first attempt included the staging of deception operations that clearly showed Germany mobilising troops to invade neighbouring Austria. However, his attempt to rouse Austria failed, and Hitler was able to annex it with relative ease in 1938.

While all-out conflict did not follow, Canaris was convinced that, if Hitler succeeded in his next objective, taking Czechoslovakia, then war would be unavoidable. Once more, Canaris tried to sabotage the operation from within. Falsifying intelligence, he told the German High Command that the Czech military force was not only far larger than had been envisaged but they were also well prepared for any invasion attempt. To assist him in his sabotage mission, he also began to recruit fellow anti-Nazi officers into the *Abwehr*. Meanwhile, plans were afoot for a new German special forces formation that specialised in infiltration and sabotage.

Despite losing the First World War, the conflict had proved to the Germans that small groups of elite soldiers, trained in guerrilla warfare, could achieve startling results. Not only had T. E. Lawrence and his band of Arabs defeated the Ottoman and German forces in Arabia through such means, Germany had also seen success in its east Africa colony. Like Lawrence, Lieutenant Colonel Paul Emil von Lettow-Vorbeck's *Schutztruppe* – 'protection force' – had successfully used guerrilla tactics, and locally recruited Askaris, to defeat much superior forces.

However, Germany had already tried to employ such tactics during the 1939 campaign in Poland with mixed results. The OKW (Armed Forces High Command) had formed a unit of Polish-speaking ethnic Germans, called *Bataillon Ebbinghaus*. While it had conducted irregular missions, such as infiltration, sabotage, securing roads, bridges and similar assets, it had also facilitated German progress by sowing confusion within enemy ranks. Yet their unorthodox methods and heavy casualties, as well as incidents involving the killing of civilians, proved too much for the Prussian High Command, who disbanded the unit soon after. Despite this, some still believed that such special service units would be vital to German success.

One such man was Dr Theodor von Hippel. Von Hippel had not only served under Lettow-Vorbeck in east Africa, but he was also a student of Colonel T. E. Lawrence and his exploits. Now von Hippel sought to emulate them by creating the first German 'special forces', who would operate behind enemy lines and specialise in psychological warfare. But his proposals were rejected by his military superiors, including Canaris. Not only did he view von Hippel with suspicion, believing him and his proposals to be pro-Bolshevik, but he also told him that psychological and ideological warfare were the prerogative of the ruling political party, not the military.

Unswervingly committed to his idea, von Hippel turned to his former commander, Helmuth Groscurth. Having long advocated the formation of ethnic German sabotage units, Groscurth strongly backed von Hippel's suggestion and brought pressure on OKW to consider the idea more carefully. With this added impetus, authorisation was soon received to create the unit, which became known as the 'Brandenburgers', due to it being formed and trained in the German state of Brandenburg.

The unit was to be placed under *Abwehr* command and, once Canaris was reconciled to this decision, he changed his stance, seeing it instead as a way to potentially enhance his own subterfuge goals. Working from the inside, he commenced a gradual but systematic removal of National Socialist officers from von Hippel's command. Fellow anti-Nazi, Erwin Heinrich Rene von Lahousen, head of the so-called *Abwehr II*, who would be overseeing the Brandenburgers, said of Canaris's goals for him:

Beginning with his very detailed conversations with me relative to me taking over Abt. II of the Abwehr (beginning of 1939) up until my departure from the Amt./Ausland Abwehr (middle of 1943), Canaris had spoken constantly – whether by paraphrases and hints or quite openly – of the necessity of doing away with Hitler, Himmler, Heydrich and of disposing of the whole criminal gang. He also explained therewith my appointment as head of Abwehr II and the real reason for forming the Brandenburg Regiment ... The role that he had destined for my Abteilung, specifically for me and the Brandenburger Regiment, was as follows: I was to prepare myself at a given time for the acquisition of material (explosives and time fuses) for the accomplishment of the 'action'. On the other hand, the Brandenburg Regiment was, to a degree, to be set up as Special Troops at the disposal of that first, powerful occupation by certain 'key units' of the National Socialist juggernaut (the RSHA, radio network, intelligence branch of the OKW etc).

Despite Canaris's ulterior motives for the Brandenburgers, he couldn't take complete control of all aspects without giving himself away. While he could recruit anti-Nazis at a high level, it would be all but impossible to do so across the whole unit, particularly with von Hippel put in charge of calling for volunteers.

While von Hippel appealed to existing Wehrmacht formations to join this burgeoning elite force, he already had a certain idea of the kind of recruit he was looking for. An aptitude for foreign languages was a prerequisite if they were to work behind enemy lines, but any candidate also had to be motivated, adventurous and physically fit and show ability in improvisation, marksmanship and self-control, as well as having a sound knowledge of foreign customs and cultures. The optimal recruit thus tended to be college-educated and had usually worked in civilian occupations before the war. This generally ensured that such a person possessed a degree of maturity, experience and self-sufficiency. As the war expanded, recruits from occupied Europe, and areas slated to be captured, were also invited to join. Moreover, with the Brandenburgers' reputation for conducting daring, dangerous missions spreading, regular army soldiers also increasingly applied to what they viewed as an elite formation.

Initial Brandenburg training took place at *Abwehr II*'s Kampfschule Quenzgut, on an estate along Lake Quenzsee, just west of Brandenburg. There the Brandenburgers learnt to exemplify their credo of 'seeing without being seen'. Of vital importance was the ability to go undercover. While they were expected to be fluent in the language and customs of the country they would infiltrate, they also studied enemy uniforms and ranks, and mastered foreign weapons and vehicles.

In terms of disguise, the Brandenburgers usually had two options. The so-called *Halbtarnung* (literally, 'half camouflage') essentially consisted of wearing an enemy greatcoat over a German uniform, with corresponding headgear. Meanwhile, *Voll Tarnung* ('full camouflage') saw an entire enemy uniform worn over the German one. The latter was obviously more convincing, but it would not be uncommon for the Brandenburgers to begin combat before they had removed their disguises. Wearing full camouflage therefore meant they ran the risk of shooting each other in the confusion of battle. Indeed, the wearing of full camouflage also had other potential consequences. The enemy would be able to declare the wearer a spy and execute them.

While going undercover was essential, so was the ability to carry out certain operations, which usually involved sabotage. As such, the Brandenburgers learnt how to destroy different targets, such as bridges, power stations, industrial and railway installations, ships, cables and wireless stations.

In effect, the Brandenburgers were James Bond-like figures, which incidentally was a role I almost came to play. In the early

1970s, I received a telegram quite out of the blue from the William Morris Agency asking me to go to London to audition for the part of James Bond. Sean Connery had retired, his successor George Lazenby had been pensioned off, and the prime mover behind the Bond films, Cubby Broccoli, was on the lookout for a new 007. Mr Broccoli, I was told over the phone, was looking for an 'English gentleman who really does these things'.

'What things?' I asked.

'Shoot rapids, climb drainpipes, parachute, kill people, you know . . .'

The fact that I couldn't act seemed irrelevant, and it was an all-expenses-paid trip to London, so I went down to try my luck. Somehow, I managed to get through all the screen tests, and in my mind was ready to be fitted for my tuxedo and drive an Aston Martin, but it was not to be. After a ten-minute interview, Mr Broccoli decided I was too young, most un-Bond-like and facially more like a farmhand than an English gentleman, a comment that certainly left me a little shaken and stirred. Yet, while the world was to miss out on my thespian talents, my cousin Ralph was to gain a recurring role in Bond films when he took over as M, the head of MI6, from Dame Judi Dench in 2012.

But, unlike Bond, the Brandenburgers were the real thing and, as part of German combat operations in the west, they participated in a variety of covert missions in 1940. In Denmark, they secured primary roads and river crossings near the border while in civilian dress, as well as via glider at the Green Belt Ridge wearing Danish uniforms. During the following month, they also assisted alpine units in Norway in securing bridges in Holland, and similar structures in Belgium.

When the Germans turned eastward in 1941, the Brandenburgers remained an important complement to traditional operations in the Balkans and Yugoslavia, and even conducted clandestine missions in areas as distant as Persia, Afghanistan, India and north Africa.

However, while the unit's first major mission was Operation Barbarossa, in which they played a crucial role in the invasion of the Soviet Union, it was their capture of the Maikop oilfields that really cemented the Brandenburgers as an elite force in Hitler's army, despite Canaris's efforts to use them for his own purposes.

By 1942, Hitler's ability to wage war was becoming repeatedly hamstrung by a lack of resources, most notably oil. Pre-war German oil supplies had originated from three sources:

1. Imported crude or petroleum products from overseas;
2. Domestic production in Germany and Austria; and
3. Synthetic oil produced domestically, primarily from coal.

During the last full year of peace, German consumption of oil totalled 44 million barrels, of which 60 per cent was derived from foreign imports. The outbreak of war not only resulted in increased demand, it also cut off supply from most foreign sources. In September 1939, German stockpiles had plummeted to 15 million barrels, with demand only going upwards. In 1941, OKW estimated that Germany's oil supply would be exhausted by August. Things were getting serious.

Romania's Ploiesti oilfields subsequently became Germany's chief oil supplier, providing almost half of the nation's entire consumption. Nonetheless, it was still below German expectations. Yet there was no sign of any available increase as the oilfields themselves were gradually becoming depleted. Hitler was therefore forced to put the seizure of oil at the top of his list of military objectives or face inevitable defeat. As such, he now turned his attention to oilfields in the Soviet Union, telling assembled officers of Army Group South in June 1942, 'If I do not get the oil of Maykop and Grozny, then I must end the war.'

With German troops already in the Soviet Union, they commenced laying the groundwork for Operation Edelweiss and the capture of the Caucasus oilfields in June 1942. Quickly clearing the route towards Maykop of Soviet resistance, they were eventually forced to stop 150 miles from their target. It appeared the Soviets were well aware of Hitler's intentions, and they had not only placed a strong concentration of troops around the oilfields but had made preparations to destroy them in an act of sabotage should the Germans get any closer. It was now time for the Brandenburgers to act. Lieutenant Adrian von Folkersam of 3rd Company was subsequently tasked with infiltrating behind Soviet lines and capturing the oilfields, or at least preventing their demolition.

Folkersam, known as 'Arik' to his friends, had been born into an aristocratic Baltic German family who had long served the imperial Russian military. However, Folkersam's family had fled Russia following the Russian Revolution in 1917 and settled in Latvia. Growing up in Riga, Folkersam went on to study economics in Munich, Konigsberg and Vienna, after which he became a journalist on the *Rigaschen Rundschau* and joined the SA. Known as a charismatic team leader, and ticking many of the required boxes, Folkersam was accepted into the Brandenburgers' 2nd Company, which comprised Germans of Baltic, Russian and African origin. Wolfgang Herfurth of SS-Jager-Bataillon 502 later described why he believed Folkersam was the ideal Brandenburger:

He was a man of few words and able to play many parts well. He could distinguish exactly what was important from the unimportant. He always had reservations about orders coming from above and each order was examined and checked for its feasibility. So, he sometimes changed orders without further approval. He was completely convinced of his family's tradition and lineage and a supporter of the elite ideals as well as a follower of Stefan George's poetry, which brought us closer to humanity. He carefully prepared his subordinate commanders with exercises and difficult question and answer contests. He put great emphasis on individual training and strengthening. And though he clearly had a great deal of confidence, he was no arrogant superior.

For his mission, Folkersam selected sixty-two men, mainly Baltic *Volksdeutsche* whose Russian was perfect, plus a scattering of hand-picked Sudeten Germans. He decided they would be in full Soviet uniform, without the usual precaution of a German uniform underneath. Given the nature of the deep penetration of enemy lines that was planned, it was deemed an unnecessary danger to carry any form of identifiable Wehrmacht equipment. They were equipped instead as NKVD troops of the 124th NKVD Rifle Brigade, with Folkersam taking the guise of Major Truchin. Such was the danger of the mission that each soldier carried a cyanide capsule to take in case of capture, as it was preferable to the torture that would inevitably follow.

Because of the rapid German advance, and corresponding retreat of the Soviet 9th Army, the Russian front line was extremely porous. This allowed Folkersam's group to march under bright moonlight, through acres of sunflower fields, without being confronted. Yet soon they came across Soviet troops at the small village of Feldmarshalskiy. This would be their first major test.

Walking down the narrow road, the Brandenburgers had no option but to join the mix of retreating Soviet troops, who had become separated from their parent units. Despite fighting for the Soviets, these men were from diverse backgrounds, including Kuban Cossacks, Ukrainians, Kyrgyz groups, Georgians, Caucasians and Turkmen, as well as Russian and Siberian units. From careful observation, Folkersam's patrols found that only the Russian and Siberian men were keen to return to the fighting. The remainder either favoured defecting to the Germans or melting away to their homes. Folkersam could sense that the few Soviet officers present were also struggling to retain control.

Reaching the village in the darkness, Folkersam saw that alongside the enemy's horses and camels were trucks and fuel supplies. He decided these would be beneficial to their mission, and having heard the conversations between the men he knew just how to take them.

Aggressively firing his gun into the air, Folkersam and his men quickly moved in to disarm the startled Soviet troops. In his NKVD uniform, Folkersam then jumped onto the bonnet of a truck and began haranguing the group, accusing them of desertion in the face of the enemy. He then ordered all Cossacks to be separated from the rest and, along with a small contingent of his force, marched them out of the village for proclaimed summary execution as traitors and cowards.

Away from the prying eyes of the remaining Red Army, Folkersam rounded on the Cossack leader and asked whether his men genuinely intended to defect to the Wehrmacht, causing confusion and suspicion among the prisoners. When it was clear that their intention was genuine, Folkersam surprised them by offering them the opportunity to head towards Anapa, where the German lines were stationed, while he and his men fired volleys into the air that would satisfy the listening men that they had indeed been executed. The deal was struck and after the Cossacks' departure and the ensuing gunfire, Folkersam and his men returned to the village. There, they ordered the Russian and Siberian officers present to head to the Soviet lines, while Folkersam's men commandeered the trucks.

The journey was long, hot and dusty, as the main road south soon became clogged with refugees and retreating troops. Reaching Armavir, on the Kuban River, the Brandenburgers were stopped by genuine NKVD troops, who were attempting to restore order among the teeming throng of people. With his trademark confidence and charm, Folkersam swung down from the truck and presented his forged credentials to the NKVD colonel in charge, claiming to be on special assignment on behalf of Aleksei Zhadov, commander of the 66th Army at Stalingrad. Unwilling to admit ignorance of the order, the colonel chastised Folkersam for being a day later than expected and directed him to continue towards Maykop, while warning him to remain vigilant against 'fascist spies' disguised as Soviet infantrymen. With a smile, Folkersam said he would be sure to do that.

Upon finally reaching Maykop, with little further incident, Folkersam drove straight into the lion's den: the Red Army headquarters. With characteristic self-assurance, he introduced himself to General Perscholl, the NKVD lieutenant general charged with local defence. To his surprise, Folkersam was warmly greeted. The news of his execution of traitorous Cossacks had already preceded his arrival and seen him heralded as a true patriot.

With this, the Brandenburgers were provided with a confiscated villa and adjacent garage for the duration of their stay. Perscholl even invited Folkersam to dinner, and over a full meal, accompanied by large quantities of vodka, the commander established himself as a welcome guest in the city. But he didn't forget his mission.

During the days that followed, Folkersam and his men were able to reconnoiter Maykop's defences. Folkersam even went as far as using his NKVD position to order anti-tank positions to be moved away from the main road, persuading Perscholl that the German panzers would take a 'less obvious' axis of attack.

By the evening of 8 August, German armoured forces were only 20 kilometres away and chaos had overtaken the Soviet troops in Maykop. Perscholl had now departed with his staff and there was widespread looting by roving bands of leaderless men. The time had come to execute the final stage of Folkersam's mission.

He proceeded to divide his men into three parties. The first and most numerous, under the command of Sergeant Landowski, was charged with heading southwest towards Neftegorsk, where they were to prevent the destruction of oil installations being prepared for demolition. They were to kill the genuine demolition parties and take their place, masquerading as Soviet sabotage troops. A second, smaller group, headed by Franz Koudele, was to sever telephone and telegraph communications between Maykop's defenders and the outside world, so as to ensure no back-up could be summoned. Folkersam was then to lead the third group, which was intended to meet two Soviet Guards brigades that had arrived from Tbilisi and Baku as reinforcements and attempt to persuade them to fall back from the town.

As German artillery fire landed in the city, the second team detonated explosives at a small artillery communications centre. The resulting explosion was taken by the Soviets to be a lucky German shell hit. Meanwhile, Folkersam forcefully persuaded defending artillery troops that the German threat lay south and had them abandon their defensive positions while he and his 'heroic' NKVD men covered their move. He next presented himself to the Soviet general commanding one of the reinforcing brigades and endeavoured to convince him that the front line now lay to the south. His opponent was initially sceptical, and suspicion began to transform into hostility, before word of the retreating artillerymen swung the issue in Folkersam's favour and the first brigade was ordered to withdraw, its movement convincing adjacent troops to also leave their positions and head south.

Meanwhile, Koudele had managed to bluff his way into the central communications centre and browbeat the officer in charge to evacuate the building and retreat to the new defensive line at Apschetousk. Once the Soviet operators had departed, Koudele's men manned the telephones and radios, redirecting Soviet troops towards Tuapse, as they relayed the disinformation that Maykop was cut off and surrounded. By noon, as panzers reached the northern outskirts of the city, the Brandenburgers evacuated the building and destroyed the Soviet equipment with grenades.

To the southeast, group one under Landowski raced into individual facilities and ordered Soviet troops stationed there to retreat while they handled the demolitions themselves as a 'rear-guard'. Unconvinced, a Soviet security officer attempted to contact army headquarters for back-up. Luckily for the Brandenburgers he was unable to get through, thanks to Koudele's team taking control of the communications.

By 9 August, the Brandenburgers had achieved all of their objectives with spectacular success. The oilfields were safe, the Soviet troops had been dispersed, and the road was now clear for the 13th Panzer Division to storm into Maykop to claim their prize. The men of Folkersam's command, still dressed in their Soviet uniforms, gingerly made their way towards approaching German troops and 'surrendered' themselves. Fortunately, the leading elements had been fully briefed about their operation and there were no accidents. Folkersam's mission was now over and for his efforts he was awarded the Knight's Cross on 14 September 1942.

While the Brandenburgers' role at Maykop allowed Hitler to stay in the war, Wilhelm Canaris was still determined to stop him. During 1943, he made direct contact with both the Americans and British in an attempt to convince them that there was serious resistance to Hitler within Germany. His suggestions to end the war included killing Hitler, or handing him over to the Allies. All were rejected. Taking matters into his own hands, he eventually tried to assassinate Hitler in 1943 and 1944, before Hitler realised that Canaris was involved in the conspiracy against him.

Canaris was subsequently arrested and transferred to Flossenburg concentration camp, where he was kept on starvation rations, regularly beaten, mocked, tortured and humiliated. However, at no time did he reveal any information that could be used against his fellow conspirators. On 9 April 1945, following a bogus trial staged by two SS officers, Canaris and several of his colleagues were stripped naked and led to the gallows. With their SS guards mocking them and hurling abuse, they were hanged and their bodies left to rot. Just two weeks later, the camp was liberated by US troops.

But, before victory could be claimed, the Allies were preparing to launch an assault on the Normandy coast that would represent the biggest amphibious invasion of all time. However, for it to succeed, the German coastal guns had to be silenced behind enemy lines or else it could turn into a massacre. Once more, Churchill turned to his special forces . . .

22

THE PARATROOPERS

By 1943, with the Americans now having joined the war against
Germany, thoughts turned to how to retake Europe from the
Nazis. Many advocated mounting an amphibious operation
on the coast of France and driving the Nazis back from there.
However, this was not without its problems. In August 1942,
a raid on the French coastal town of Dieppe had already high-
lighted the huge issues involved in attacking an enemy-held
port. The German beach defences had kept the invading troops
confined to the water's edge while machine-gun fire then decimated them as they lay trapped in the open. In light of this, if
any beach landings were going to be made in future, the German
defences had to be taken out beforehand. But how could this be
achieved? Thankfully, the Germans were to provide their enemy
with the inspiration to embark on one of the most important
raids of all time.

In 1935, as part of manoeuvres in the Kiev military district,
the Soviet Red Army invited foreign military attachés to watch
a force of 1,200 military parachutists jump from a number of
Tupolev aircraft. The visitors were subsequently treated to the
spectacle of masses of parachutists tumbling off the wings of the

large six-engine transports and landing in the fields around them. However, while the watching British guests believed the act to be nothing more than a novelty, the Germans were captivated. They instantly grasped the potential of using speed and surprise to defeat an enemy by attacking him in a manner and location least expected.

By the outbreak of war in 1939, the Germans had a complete parachute division of 9,000 *Fallschirmjäger* and a separate division of troops who could land by glider or transport aircraft. In 1940, they would embark on their first major mission, when small numbers of parachute troops were dropped to capture strategic targets in Norway and Denmark. While success was mixed, it was in the Low Countries shortly afterwards that their ability would really be demonstrated.

The heavily defended Belgian fortress of Eben-Emael occupied a strategic point on the Dutch–Belgian border, with its 120mm and 75mm guns dominating three bridges over the Albert Canal. These bridges were deemed critical to the success of the German invasion, but the fortress was considered impregnable. On the night of 10 May 1940, seventy-eight *Fallschirmjäger* achieved what many felt to be impossible.

Landing silently by glider on the flat grass-topped roof, the small force used shaped charges to cripple the guns and blasted their way inside. Before the garrison could react, they were trapped. After thirty hours of fighting, the garrison of 600 was forced to surrender, having suffered over 100 casualties. Losses to the much smaller German force amounted to six men killed and nineteen wounded. Meanwhile, other troops landing by parachute had captured two of the three bridges intact.

To their north, the Germans used the rest of their airborne forces to capture strategically important bridges over the Maas, which led to the rapid capitulation of Rotterdam and the defeat of Holland. This spectacular feat confirmed Hitler's faith in his paratroopers' effectiveness; it also stirred the imagination of Britain's new prime minister.

On 22 June 1940, Churchill wrote a memorandum to the chiefs of staff and directed them to raise a corps of at least 5,000 parachutists. This was a tall order. It had taken Germany four years of trial and error to get its paratrooper regiments to reach this high standard. Britain was now trying to match them within a few months. Unsurprisingly, not everyone at the War Office or Air Ministry was convinced of such an idea.

Conventional thinking in the senior echelons of the military believed all efforts should be directed towards preparing the army to meet the imminent threat of a German invasion, while also building up the strength of the RAF to defend the country's skies. They also viewed the success of German airborne operations in Europe as a one-off. But still, the prime minister had issued his directive and it needed to be satisfied.

As such, the War Office distributed army circulars to units calling for volunteers. To secure sufficient numbers of volunteers, recruiting teams also visited units throughout the UK. Many men jumped at the opportunity, quite literally. The Parachute Regiment seemed to offer an exciting diversion from home defence duties, with a greater opportunity for making an appreciable contribution to the British war effort. The extra pay – two shillings per man and four shilling per officer per day – was a further incentive.

All prospective candidates initially underwent a rigorous selection process and instruction at the Airborne Forces Depot at Hardwick Hall, near Chesterfield in Derbyshire. Everyone was required to undergo a strict medical, which weeded out all those who had previously broken a leg, had false teeth or needed spectacles. A test to determine whether men were susceptible to airsickness was also carried out by placing candidates on a stretcher suspended from the roof and vigorously swinging it through the air for twenty minutes. As you might be able to tell, the recruitment methods were still relatively crude but much of it was still trial and error in these early days.

For those who passed these initial tests, army physical training instructors then put candidates through a gruelling course lasting a fortnight. This included intensive physical gym tests, assault courses and repeated road and cross-country marches throughout the surrounding area, while carrying full arms and equipment. This tough course ended with seven difficult tests, including running 2 miles in sixteen minutes in full battle order. Only those judged as having sufficient stamina, self-discipline and physical and mental resilience passed the selection, with many potential recruits – normally the majority – 'Returned to Unit'.

But just passing through Hardwick Hall didn't necessarily mean that a candidate had been accepted. The real acid test was still to come.

Those who passed this first stage then had to undertake a fortnight's parachute training at the Central Landing School, located nearby at the former civil Ringway airport near Manchester. As the idea of a paratrooper unit was so new, the army and RAF instructors had to learn the basics of parachuting themselves. Moreover, suitable parachutes, equipment and techniques for jumping from aircraft also had to be developed. As military historian Julian Thompson has written, 'Despite the total lack of practical knowledge of parachuting techniques, text books, special equipment, and in some cases, the basic inability to impart knowledge to others, the RAF and Army instructors of all ranks shared a boundless enthusiasm and great courage.'

Quickly putting together a 'curriculum', the parachute phase of training began with a week of 'synthetic' ground training. This saw trainees learn how to exit from different types of aircraft, and then how to control their parachutes in the air during descent and landing. Mock-ups of various planes' fuselages were subsequently constructed in hangars and these were used by recruits to practise dummy jumps. Other types of training apparatus helped men learn flight drills and correct methods of landing. These included trapezes, from which recruits were suspended, and wooden chutes, down which men plummeted and then dropped from some height, learning to fall and roll, an important skill to prevent injury.

At the end of the first week, recruits made two descents from a tethered, static barrage balloon at a height of 700ft. For many, this provided their first terrifying experience of parachuting to the ground. It was not a popular stage of training, with many extremely nervous men suffering bouts of acute nausea while ascending in the balloon as it was buffeted to and fro in the wind. From a hole in the middle of the balloon's basket, men then fell a distance of 120ft before a static line made the parachute snap open. This actually sounds more terrifying than jumping from an actual plane.

During the second week at Ringway airport, trainees progressed to making a series of parachute descents from aircraft. These were in fact converted bombers, with holes cut in the floor. Casualties were not uncommon. For instance, on 25 July 1940, Private Evans plummeted to his death when the rigging lines of his parachute twisted round his canopy.

I once had a very similar experience when preparing to join the SAS. Attending a parachute school at Pau, near Lourdes, I was already very apprehensive due to my fear of heights. My worst fears weren't helped when a fellow student explained why the school was near the holy city of Lourdes. 'You see, those men who are crippled here at Pau,' he helpfully explained, 'they can seek recovery in the holy waters just up the road.' Shortly afterwards I embarked on my first jump from a Nord Atlas transport plane. Closing my eyes, I threw myself out before I had the chance to become afraid. However, a few seconds after my parachute had apparently successfully opened, I looked up to see a knotted tangle, which met in a bunch around my neck. I experienced instant panic. The ground was fast becoming uncomfortably close.

In fact, I was merely experiencing a common problem called 'twists', usually caused by a poor exit from the aircraft. The normal process of elasticity slowly unwound the tangle, but I was spinning like a top on landing and hit the ground with a wallop. I would count myself fortunate to escape with just a few bumps and bruises. A few years later, when embarking on an expedition to explore Norway's Fabergstolsbre Glacier, I would again have to jump out of a plane. Once more all did not go well. Upon jumping out of the Cessna, I almost hit the fuselage and fell into a body spin. As I did so, a camera came loose in my anorak and lodged itself against the ripcord bar, preventing me from opening my chute. In a blind panic, I frantically grasped at the camera, as the freezing air blasted my face and I plummeted towards the solid ice down below. With seconds to go, it came loose and allowed me to open the chute, and I watched with relief as the orange canopy now fluttered above my head.

However, at Ringway, many were not as fortunate as I was. By the end of 1940, at least three men died during the first 2,000 descents. Such was the terrifying nature of this training that, in 1940–41, over half of the candidates failed to pass the two-week course. For those who did, they received their coveted parachute wings and distinctive maroon berets. Incidentally, this happened to be the favourite colour of the novelist Daphne du Maurier, the wife of Brigadier Frederick Browning, who was the commander of the airborne forces.

When the newly qualified paratroopers joined their battalions, they then needed to learn the skills required of airborne light infantry. It wasn't enough for them to just fall out of planes. Such airborne soldiers would normally be fighting against superior odds, and against an enemy equipped with heavy weapons, artillery and tanks – all of which they would lack.

Cultivating the right attitude was seen as something of the utmost importance. This included the development of courage, aggressiveness, self-discipline, self-reliance, and, in turn, a high level of esprit de corps within each unit. Particular attention was paid to developing and maintaining a high standard of individual military skills – marksmanship, skill-at-arms and fieldcraft. A high degree of physical fitness was also deemed essential. Recruits spent much time on seemingly endless route marches and repeated assault courses, carrying full arms and equipment. Indeed, the ability to cover long distances at high speed became a particular matter of pride in the Parachute Regiment.

Exercises were also held to replicate potential operations. These usually involved descending by parachute to a target, followed by intensive tactical training, often using live ammunition, with a particular objective in mind, such as establishing bridgeheads by capturing road and railway bridges and coastal fortifications. Invariably, they ended with a long march back to base at a considerable pace.

On 31 May 1941, with the initial batch of recruits ready, the military chiefs of staff issued a joint paper outlining plans to raise two parachute brigades and an airlanding brigade, as well as to start production of troop-carrying gliders. In December, the War Office announced the formation of the Army Air Corps as the parent regiment. This would include the newly formed Glider Pilot Regiment of army pilots trained by the RAF to fly the wooden Horsa gliders that were beginning to come off the production lines. These could carry a pilot and a co-pilot, plus twenty-eight fully armed men, or two Jeeps, or a 75mm howitzer or a quarter-ton truck. Most who saw the glider described it as a 'big black crow'. The first forty candidates would begin training at the RAF Elementary Flying School at Haddenham near Oxford the following month. Within two years, the strength of the trained glider pilots in the new regiment was to grow to over 2,500.

While Churchill's plan to establish a paratrooper unit had initially been derided by many in the War Office, Britain's airborne forces would soon come to play a key part in winning the war. In March 1943, Lieutenant General Frederick Morgan was appointed COSSAC – chief of staff to the supreme allied commander (designate) – and charged with planning the cross-Channel invasion, and the liberation of Europe, taking into account all of the problems that had been raised during the failed raid on Dieppe in 1942. In studying the possible sites for the assault, it soon became clear that the area of Normandy would be the most suitable location for the amphibious forces to come ashore.

Five Normandy beaches were subsequently designated for the assault, code-named 'Utah', 'Omaha', 'Gold', 'Juno' and 'Sword'. It was decided that Sword Beach, which stretched from St Aubin-sur-Mer in the west to the mouth of the River Orne in the east, would be targeted by British forces. Eight miles up the Orne was the city of Caen, and from the city a network of roads linked it to all parts of Normandy. Taking Caen was therefore seen to be of huge strategic importance.

However, it was clear that there would be two main obstacles to this advance. The German gun battery at Merville could decimate any British seaborne landings and end the attempted invasion before it had even reached the shore. Meanwhile, securing the bridges over the River Orne and Caen Canal before the landings would be crucial. They not only provided the most direct route to Caen, but they would also enable Rommel's panzer divisions to cross and head off the left flank of the British invasion force on Sword Beach. And if Rommel's tanks were to do this, they might very well roll up the entire invading force, division by division.

It was therefore imperative that the guns be silenced, and the bridges taken intact, before any such invasion could even begin. Suddenly, all eyes turned to the paratroopers.

Neither task would be easy. Not only would the bridges and the battery be under guard, but intelligence suggested that the bridges were fitted with explosives. Should there be any attempt to take them, then the German guards were ordered to blow them up. Yet both objectives had to be completed if the Allied invasion of Europe was to get underway.

As such, glider-borne D Company, led by Major John Howard, was selected to secure the bridges. Its orders read:

> Your task is to seize intact the bridges over the River Orne and canal at Benouville and Ranville, and to hold them until relief . . . The capture of the Bridges will be a coup de main operation depending largely on surprise, speed and dash for success.

The operation was code-named Deadstick and had been designed by General Richard Gale, who commanded the 6th Airborne Division. Just a few hours before the invasion commenced, the Horsa gliders would fly into France and set twenty-eight fighting men beside each bridge simultaneously. For success to be achieved, the gliders would have to arrive like thieves in the night, without noise or light, unseen and unheard.

With preparations for Operation Deadstick underway, Gale now turned his attention to the Merville gun battery. He looked to the 9th Battalion of the Parachute Regiment, for a mission code-named Operation Tonga.

However, destroying the guns was a difficult task. They were protected by 2-metre-thick concrete walls and overhead cover, with another 2 metres of earth banked up at the sides and on top of the structure. The guns and their crew could also be secured from the outside world by heavy steel doors to the rear and steel shutters over the gunports. An air-filtration system was also installed to guard against a gas attack. In short, the bunker was strong enough to withstand anything but a direct hit from the heaviest of Allied bombs. The operation would require extensive intelligence, and intensive training, if it were to stand any chance of success.

As such, preparations for Operation Deadstick and Operation Tonga were aided by detailed overhead photographs, as well as information gleaned from the French resistance. This allowed for models of the bridges and the guns to be built to assist with planning. Lieutenant Henry Sweeney said of this treasure trove of intelligence: 'I don't think there was a bush we didn't know the height of, a ditch we didn't know the depth of or whether it had any water in.'

Intelligence indicated that the garrison of the two bridges consisted of about fifty men, armed with four to six light machine guns, one or two anti-tank guns and a heavy machine gun. While it was believed that the bridges had been prepared for demolition, resistance intelligence had pinpointed the pillbox where the charges were contained. If the British could get to the pillbox before the Germans, then they could save the bridges. But, to achieve this, they would have to move with real speed and stealth.

However, the photographs revealed a problem. Near the bridges, small holes were being dug across the glider landing sites in order to erect anti-glider poles. These were 8–10ft high and often enhanced by tripwires and topped with explosives. If they were erected before D-Day, then this represented a significant hazard to Operation Deadstick. Getting the gliders down undetected, in the dark and in a precise location, was already hazardous enough.

There were also issues for Operation Tonga. The Merville guns were not only defended by a garrison estimated to be in the region of 160–180 men, with dug-in machine-gun pits, but they were also surrounded by extensive minefields and a coiled barbed-wire perimeter several metres thick. There was also the difficulty of the surrounding area being marshy with a large number of drainage ditches. If a paratrooper should be unlucky enough to land in one, then he would drown due to the weight of his equipment, with each man estimated to weigh around 250lbs.

Still, despite these significant hurdles, preparations continued and orders were finalised. D Company was tasked with securing the bridges, intact, within minutes of landing. Meanwhile, the 9th Battalion had to destroy the Merville guns before the coming of first light at 0525 hours. Upon achieving this, they were to send the codeword 'Hammer' to HMS *Arethusa*. If they were unsuccessful, they would send the codeword 'Hugh'. If the *Arethusa* heard this codeword, or did not hear from the paratroopers by 0525, its orders were to fire at the guns, in a last desperate effort to secure a direct hit, before the troops landed on Sword Beach. Should communications fail, there was also a back-up signal: yellow smoke to denote success, while a carrier pigeon would be on hand to send word if all else failed.

D Company, the 9th Battalion and their respective reinforcements were soon training under the strictest secrecy for what would be the most vital operation of the war. Failure was not an option. Not only did the lives of thousands of British troops depend on them, so did the fate of Europe. They trained repeatedly in all conditions, learning to master every weapon that might be available to them, including German ones, and going over their respective operations again and again, in locations that were made to resemble their targets as much as possible.

Billy Gray, who was part of Operation Deadstick, recalled:

We knew exactly what we had to do. We trained and practised it so often that we knew it like the back of our hand. Anyone could have taken each other's place. Each individual soldier knew exactly what he was supposed to do on the night ... I don't think there was an incident that could have happened that we hadn't rehearsed in one way or another.

After months of preparation, the go-ahead for Operation Deadstick and Operation Tonga was finally given on 5 June 1944. Operation Deadstick, and the troops of D Company, would be the first to get underway.

In the evening, D Company gathered in front of Major Howard for a last-minute pep talk before Private Wally Parr chalked 'Lady Irene', the name of his wife, onto the side of the glider in which he would be travelling. As they boarded the gliders, the men looked a fearsome sight. With blackened faces, some had also shaved their hair to the bone. They were also armed to the max, each carrying a rifle, Sten gun or Bren gun, six to nine grenades and four Bren-gun magazines, while some had mortars.

At 2256 hours, D Company took off in 'Lady Irene', with other gliders following behind at one-minute intervals, all being tugged by Halifax bombers. In the back of these 'black crows', everyone sat tensely, squashed together. Some vomited due to airsickness, or fear, while others said silent prayers. Others sang songs to gee up themselves and their comrades. Yet, no matter how each man dealt with the situation, all were aware that this was a highly dangerous mission and many of them would soon be dead.

Whenever I would feel afraid of death on my expeditions, I would remember that I was not alone in facing the dreaded hours ahead. I would picture my grandfather who had trapped in northern Canada and fought for his country all over the world. I thought of my father and uncle who were killed in two world wars. I pictured my wife, my mother and my sisters and I knew that all of them were right behind me. I have no doubt that many of these men would have been having very similar thoughts of their own families, and would have looked at their pictures in their hands.

Suddenly, just after midnight, the crashing waves of the Normandy coast came into view. With this, 'Lady Irene' was cast off. However, as the glider silently drifted into France the clouds covered the moon, so that the bridges and the glider landing zones could no longer be seen. All the pilot, Jim Wallwork, had to direct him was a watch and a compass. This was hard enough at the best of times, especially as he still wasn't sure if the anti-glider poles had been erected. Moreover, Wallwork had also been given the impossible task of taking out a barbed-wire fence as he landed, making it easier, and quicker, for the troops to reach the bridges, with every second vital if they were to take them intact.

At 0014, Wallwork could suddenly make out the grey form of Benouville Bridge up ahead. As he brought his glider down, it zipped past trees at 90mph, bounced off the ground on impact, then ploughed through the barbed-wire fence, destroying its nose as it did so. Wallwork and his co-pilot were thrown from the cockpit while the troops inside were all knocked unconscious. With every second crucial, they needed to wake up urgently before they were discovered.

As they slowly came round, they found that despite the bumpy ride they had not only put down in the designated landing zone, with the barbed-wire fence taken out, they had also remained undetected. With glider two soon landing close by, now was the time for action.

Bundling out of their gliders they raced to the canal bridge in their camouflaged battle smocks, holding their guns at their hips. On seeing them, the few Germans on the bridge quickly retreated and screamed out to their colleagues in the pillbox. There wasn't a second to lose.

As the Germans in the pillbox scrambled to react, the British soldiers ran with all they had towards it before throwing grenades inside. An explosion followed, along with great clouds of dust. The bridge was still intact but the pillbox had been destroyed.

Not taking any chances D Company searched the bridge for explosives, cutting all fuses and wires, only to find that they had not actually been fitted. The Germans had not foreseen an airborne attack and had believed that, if the Allies should attack via the ocean, then they would have enough time to fit them before they landed. It was to prove a fatal mistake.

Soon the cry went out, 'Ham and Jam! Ham and Jam!', to signify the mission's initial success. By 0040 hours, having overwhelmed the German platoon of the 736th Infantry Regiment, the bridge and the second structure over the River Orne 600 metres away had been secured.

Now they just had to hold both bridges from any German counterattacks before they would be relieved by the commandos in the morning, provided they managed to land successfully at Sword Beach. This aspect was now in the hands of the 9th Battalion, who had to ensure the guns at Merville were taken out beforehand.

At that very moment, thirty-two Dakotas were in the sky, each carrying twenty men of the 9th Battalion. Each had a luminous skull and crossbones symbol painted on the left breast of his smock to avoid being shot by his own side. Below them in the Channel, from the Solent to the far side of the Isle of Wight, the water was crammed with ships in readiness for the D-Day invasion. It was the biggest naval force ever assembled for war and a reminder of the importance of their mission.

Before the Dakotas discharged the 9th Battalion over their drop zones, 100 RAF Halifax and Lancaster bombers unleashed 4,000lb 'cookies' on the battery at 0030 hours. From the coast, the fleet of Dakotas could see the orange explosions light up the landscape. Soon the cry went out, 'Get ready! Stand up! Hook up! Check Equipment!' The paratroopers responded instantly, clipping the snaplinks of their static lines onto the inboard cable running the length of the roof of the aircraft. All did a last-minute equipment check as they anxiously waited for the red bulb to turn green, to signify they could jump.

Suddenly, red tracer fire from German anti-aircraft guns lit up the sky. The Dakotas' tight V formation disintegrated as they tried to avoid the incoming fire, with the paratroopers thrown around inside like ragdolls.

In the darkness, and disorientated from coming under attack, some of the pilots lost their bearings and could not see their designated drop zone. Like the glider pilots, all they could do was use time, speed and distance calculations to work out their positions and hope they were as close as possible. When they estimated they were over their drop zone, the pilots flicked the switch to turn the cabin bulb green, with the dispatcher yelling, 'GO! GO! GO!' One after the other the paratroopers descended into the night sky, pulling their cords to feel the reassuring tug at their shoulders and hear the crack of silk as their chutes opened and deployed into fully functioning canopies above them. However, the Germans had now seen they were coming and, as they sailed to earth, many were sitting ducks.

With the Germans opening fire, many men were killed before they had even reached the ground. Those who survived being shot in mid-air soon found that their Dakota pilots had drastically missed their drop zone. Some landed in the forest, becoming impaled on branches, while others smashed into the sides of buildings. The speed at which they left their plane, along with the gusts of wind, saw some hit the ground too fast and break their legs upon impact. The unluckiest of all found themselves submerged in the marshes or rivers and drowned under the weight of their equipment. Those who made it in one piece found that they had fallen disastrously short of their target, which led to the battalion being scattered over an area of 20 miles. It was a disastrous start to the operation, with 9th Battalion in total disarray.

Despite this inauspicious start, they needed to gather their wits quickly if they were to reach the rendezvous point on time and take the guns. In the darkness, pockets of men stumbled their way through the countryside, while trying to avoid German patrols in the process. The orders of their commanding officer were ingrained into every man: 'No person will return fire unless directly attacked by the enemy at close quarters.' Getting to the rendezvous remained paramount and there were to be no private battles. Time was of the essence, with the clock ever ticking towards the 0525 hours deadline.

By 0130 hours, some of the 9th Battalion began to assemble at the rendezvous, but it was a pitiful sight. Some were badly wounded, while none of the gliders carrying Jeeps, engineering stores, medical equipment or anti-tank guns appeared to have landed anywhere near. Only one of the Vickers medium machine guns had been recovered from the equipment containers that had been widely scattered by the Dakotas, and none of the 3-inch mortars could be found. In addition to other missing bits of vital equipment, there was no sign of the naval gunfire party and their radios, which meant that the battalion would have no means of signalling HMS *Arethusa* to confirm whether the attack on the battery had been a success or failure.

Trying to take the battery in such circumstances seemed madness. But the men had no choice. The lives of thousands of British soldiers were relying on them to make it happen.

Lieutenant Colonel Terence Otway, the commanding officer, waited for as many men to arrive as possible before having no choice but to set off. By now, it was 0245 hours and just 150 out of a total of 640 men who had jumped were present. Gliders carrying further troops were due to land on the battery, promising to increase the ranks, but the men of 9th Battalion were not to know that one of the gliders had already crash-landed in England.

Moving through the hedgerows, the remains of the battalion made their way towards the battery. Reaching the outer perimeter, a small team cut the cattle wire and crawled on their hands and knees into the surrounding minefield so as to clear a path. Progress was slow, as without mine detectors they had to feel with their hands for anti-personnel mines and tripwires. When they found a wire, they cut the strands and moved forward cautiously. It was vital no one set one off. Not only would it be fatal for the man involved but it would also inform the German garrison that an attack was in progress.

As the men crawled through the minefield, they saw some of the Germans outside the battery smoking and talking. They were clearly unaware that British paratroopers were making their way towards them. For the paratroopers, the soldiers' presence outside the battery was a good sign. The battery was not yet under lockdown.

But suddenly the Germans burst into urgent chatter. It seemed that one of the British gliders had crash-landed nearby, with all the occupants burnt to death. Inside, the Germans had found explosives, flamethrowers and pneumatic drills. Now they were well aware that they were under attack, while yet more much-needed British reinforcements had been lost. The odds against success for the paratroopers were rapidly increasing.

It was now approaching 0430 hours. The remaining gliders had still not appeared and it was starting to get light. With no time to waste, the breaching party placed their Bangalore torpedoes in the thick-coiled wire of the inner perimeter but were spotted by the now alert German garrison. With the enemy machine guns opening fire, the paratroopers raced for cover. Hiding behind trees, and in ditches, they suddenly saw a glider appear overhead, carrying vital reinforcements. But, just as it swooped low to land, an anti-aircraft tracer flashed through the sky and made a direct hit, forcing it to crash-land in the village of Merville. Soon another glider was under attack, with troops being shot as they sat in their seats and the aircraft set alight.

However, at 0445 hours, the air suddenly split with two almighty cracks. The Bangalore torpedoes had detonated and, as the explosives cleared a path, the sky filled with smoke and dust. Soon the remainder of the 9th Battalion were streaming towards the battery under heavy attack, racing across the minefield, setting off the explosives and getting torn apart as they did so.

As men fell to bullets and mines, some managed to reach the battery intact. To their surprise, they found that the steel doors at the back were partially open. Stuffing Mills bombs and sixty-six phosphorus grenades through the gaps, the sudden explosion, followed by the gas, saw the Germans come rushing out of the steel doors, gasping for breath. Now the way was clear for the paratroopers to get inside.

However, the equipment they needed to ensure the guns were destroyed had not arrived. They had to make do with stuffing them with plastic explosives and removing key parts, hoping to damage them beyond repair. By now, heavy German reinforcements were on their way to retake the battery. Meanwhile, the crew of HMS *Arethusa* had still not heard the codeword 'Hammer' to signify the guns had been taken. Having no choice, the ship was now beginning to turn its guns towards the battery, with the men of the 9th Battalion still in the area.

Time was pressing. The 9th Battalion needed to get away from the battery as soon as possible while also alerting the *Arethusa* that the mission had been a success. Without a radio link to the ship, they lit yellow candle flares in the hope that they would be spotted by Allied reconnaissance aircraft. A carrier pigeon, which had been transported in a cardboard container in the smock of a signals officer, was also released with a message that the battery had been taken.

Having done all that they could, the survivors tried to help their wounded colleagues to safety before they were all blown to smithereens. As they did so, the German reinforcements arrived, preventing them from escaping an area they thought might soon be blasted sky-high by their own ship. Engaging in hand-to-hand combat, darting behind trees and in ditches, the men did all they could to get as far away from the gun battery as possible.

Soon they heard the sound of guns erupting on Sword Beach. The largest armada ever assembled, nearly 6,000 ships, now lay off the Normandy coast. As the big guns from the warships pounded the beaches, landing craft moved forward, towards the coastline, carrying the first of 127,000 soldiers who would cross the beaches that day. Overhead, nearly 5,000 planes of all types, the largest air force ever put together, provided cover. The men looked to the Merville gun battery. It was silent. Despite their heavy losses, with just seventy-five men left from the 150 who commenced the attack, they took some comfort in knowing that they had achieved their mission against all conceivable odds. And it appeared their message had reached HMS *Arethusa*, whose guns did not open fire on them.

While the 9th Battalion had somehow achieved the impossible, D Company had spent the night holding the bridges, while coming under sustained attack. They had, however, managed to destroy the first tank that came their way with a PIAT gun, resulting in the tank blocking a T-junction, thus preventing movement between Benouville and Le Port and between Caen and the coast. It also sent a message to the German commanders that the British were present in great strength, which of course was not the case. Nevertheless, this bought D Company vital time.

As they continued to come under attack from German fighter-bombers, frogmen and soldiers, D Company saw them all off one by one until the commandos finally relieved them just after 1300 hours. To the men of D Company, this was nothing less than the arrival of the cavalry. 'Everybody threw their rifles down,' Sergeant Thornton reminisced, 'and kissed and hugged each other, and I saw men with tears rolling down their cheeks. I did honestly. Probably I was the same. Oh dear, celebrations I shall never forget.'

The road to Caen was now open to the Allies, while the Germans were unable to get their tanks to the beaches.

Following these incredible missions, the surviving paratroopers had expected to be sent back home. However, most stayed in France for the weeks and months ahead, fighting as regular infantrymen and helping British forces finally take Caen in August and, eventually, destroy the German army, leading to the end of the Second World War. Without the incredible deeds of the paratroopers, it is debatable whether D-Day could have succeeded.

With the war soon to be over, Britain would still require the use of its special forces in the years to come. Operations in the Middle East would in time inspire the creation of one of the most elite special forces the world has ever known. And when London came under terrorist attack, it would be ready ...

23

THE SAS

At 11.25 a.m., six Iranian revolutionaries, with scarves pulled tightly around their heads, ascended the steps of the Iranian Embassy at 16 Princes Gate in London. Their leader, Oan Ali Mohammed, stormed towards PC Trevor Lock, screaming, 'Don't move! Don't Move!' as he fired a deafening burst of machine-gun fire into the air.

Soon a major incident was underway, with flashing blue lights surrounding the embassy. On their arrival, the police found that the terrorists had taken twenty-six hostages and had a list of demands, these included: 'our human and legitimate rights'; autonomy for 'Arabistan'; the release of ninety-one Arabs being held in Iranian jails and safe passage for them out of Iran to the destination of their choice. Oan then stated that, if these demands were not met by noon on Thursday, 1 May, the embassy and all of its occupants would be blown up.

At this threat, the area was cordoned off, police marksmen took up positions in surrounding buildings and hostage nego-tiators attempted to establish contact with Oan. The press also flocked to the scene and soon the world was gripped by live reports of the hostage crisis. I can actually remember hearing of this incident while trudging through subzero temperatures during my Transglobe Expedition. In our communications camp, Ginny would listen to the BBC World Service and then radio us through the headlines. On that day, I recall her telling us that terrorists had stormed the Iranian Embassy. Like millions of others, I was shocked at the news, but I must confess it wasn't on my mind for too long as I still had a good few hundred miles of ice-cold, treacherous terrain in front of me.

Yet, while I braved a windchill of minus 20, it did cross my mind that an agency I had once been a part of would no doubt be preparing itself, for this was just the type of situation in which they excelled. They were the SAS and this operation would soon see their name known throughout the world.

The SAS had initially been formed during the Second World War, thanks to the efforts of a former commando by the name of David Stirling. In July 1941, while on duty in the Middle East, Stirling had grown tired of the repeated failure to take out Axis aircraft by conventional means. Rather than try to destroy them while they were airborne, he realised that a more effective strategy would be for a special forces unit to parachute behind enemy lines and hit them while they were still on the ground. But no one seemed willing to listen to his idea. Stirling therefore took matters into his own hands.

In the uniform of a junior Scots Guards officer, Stirling approached the checkpoint outside Middle East Command headquarters in Cairo. Stopped by a guard, he attempted to bluff his way inside, but without success. Yet, when the guard's attention was momentarily distracted by the arrival of a staff car, Stirling slipped past him and into the building.

Once inside the building, he entered the office of a major on the adjutant general's staff and identified himself. Stirling said he needed to speak urgently with the commander-in-chief, General Sir Claude Auchinleck. The major was unimpressed. He recognised

Stirling as an officer who had fallen asleep during one of his lectures on tactics. In his eyes, Stirling was clearly not worthy of his time and he refused to listen to his request. When the telephone then rang, with the guard reporting the intruder, Stirling knew he was about to be turfed out. Quickly slipping out of the office, he emerged into the corridor and entered the next room he saw.

The room belonged to the deputy chief of staff, Major General Neil Ritchie. Stirling apologised for his unorthodox arrival and told the general he had matters of great operational importance to discuss. Impressed by the young man despite, or perhaps because of, his daring approach, the senior officer invited him to sit down, whereupon Stirling outlined his plan.

General Ritchie listened attentively and with growing interest. The concept appeared sound and the young officer seemed convinced he could carry it off. The general subsequently approved the scheme and gave permission for Stirling to raise his force. It would be known as L Detachment, Special Air Service Brigade, and Stirling was to be promoted to captain of this new unit, where he was to prove his worth.

By the end of 1941, the SAS had claimed nearly 100 enemy aircraft – all destroyed on the ground – plus a number of vehicles and petrol stores. They had also adopted a new motto: 'Who Dares Wins'. However, Stirling was soon to be captured in Tunisia and, after several failed escape attempts, he spent the rest of the war in Colditz.

In Stirling's absence, the SAS continued to undertake breathtaking deeds of audacity in the desert, and further afield. Following the war, the SAS remained a permanent unit and evolved to take on a counter-revolutionary function, operating both at home and abroad.

While often shrouded in secrecy, the 'Land Operations Manual', a publication of the Ministry of Defence, spelled out its principal functions in 1969:

SAS squadrons are particularly suited, trained and equipped for counter-revolutionary operations. Small parties may be infiltrated or dropped by parachute, including free fall, to avoid a long approach through enemy dominated areas, in order to carry out the following tasks:

a. The collection of information on the location and movement of insurgent forces.
b. The ambush and harassment of insurgents.
c. The infiltration of sabotage, assassination and demolition parties into insurgent held areas.
d. Border surveillance.
e. Limited community relations.

Liaison with, and organisation, training and control of friendly guerrilla forces operating against the common enemy.

Indeed, in the mid-1960s, I also applied to join the SAS following my stint commanding tanks in the Royal Scot Greys. Along with 124 other hopefuls, I arrived at SAS headquarters in Hereford for a series of selection tests. The first night let us know exactly what we were in for as, in temperatures below freezing, we had to swim across the River Wye. What followed was a number of timed tests involving map reading and fast cross-country movement, all with packs weighing over 30lbs. For the first few days, we had little sleep or rest. Just as we would settle into our beds, an officer would storm in and give us another 'mission'. This saw us embark on 15-mile marches in the wet night, and then come back to face an interrogation involving a series of complex military problems. SAS staff with binoculars watched us like hawks so there was never any chance of cheating. Those who did try to take the easy way out were soon booted off the course.

I proceeded to cause a minor scandal when, after being tasked with drawing up plans to raid a bank, I accidentally left mine behind at a restaurant in Hereford. Believing it to be the real deal, the police were soon called and before I knew it newspapers all around the country covered the story. Yet, as they dug deeper, they came to realise that the plans were just part of SAS training, which saw the service crucified. In a particularly scathing article, *The Times* wrote, 'The services are letting their zeal outrun their discretion.' However, after being found to be the culprit, rather than be thrown out, I was merely given a slap on the wrist. I believe my SAS adjutant thought I had left the plans behind on purpose and found the whole thing rather funny. If he had known I had actually dropped the plans by accident, I am sure my time with the SAS would have been cut short.

With more and more hopefuls dropping out, I was soon one of the last men standing. All that was left to complete was a test known as the 'Long Drag'. This was a 45-mile cross-country bash carrying a 50lb pack, 12lb kit and 18lb rifle without a sling. I am ashamed to say that I proceeded to hire the services of a local farmer, who had a black Ford Anglia, to take me around a large part of the course and beat the clock. I did feel guilty afterwards, but I consoled myself with the thought that subterfuge was an SAS tactic and I had not been caught, so I probably deserved my place, becoming the youngest captain in the army at that time in the process. Sadly, as I have previously explained, it was not to last. Soon after my failed attempt to bomb the *Dr Dolittle* movie set, I was thrown out in disgrace.

However, I would go on to complete the Long Drag fairly and squarely a few years later. On my return from Dhofar, I discovered there was a curious anomaly known as Reserve Squadron, 22nd SAS Regiment, whose role was to provide reinforcements for the regular SAS in time of war. Applicants had to pass the regular SAS selection course and this, therefore, offered me the chance to banish the ghost. This time, I'm happy to say I passed the course without the help of my farmer friend.

However, the terrorist attack at the 1972 Munich Olympics soon changed everything. With security relatively lax, a group of eight Palestinian terrorists, members of the Black September Organisation, were able to seize the dormitory occupied by Israeli athletes, killing two and taking nine hostages. In return, the terrorists demanded the release of 200 Palestinians imprisoned in Israel. The Israelis flatly refused, but the West German government agreed to allow the gunmen, together with the hostages, safe passage out of the country. Matters quickly unravelled at the airport, where German security forces opened fire and, in the fighting that ensued, all nine remaining hostages, five terrorists and one policeman were killed.

Hundreds of millions of people watched the event on television, which not only embarrassed the West German government but also alerted other countries to the need to form counter-terrorist units to cope with similar episodes that might arise in the future. At the G7 summit the following year, western governments reached an agreement to establish forces specially trained in counter-terrorism – not least because most countries had no military personnel trained to cope with a scenario like that at Munich.

Britain was no different. While the Metropolitan Police could be employed for such a task, its skills in this realm were limited. As such, in 1973, the Counter Revolutionary Warfare (CRW) wing of the SAS was expanded, with its responsibility now including that of being the nation's hostage-rescue unit.

The CRW soon proved its worth in a string of hostage situations, most notably in Mogadishu, Somalia, where Palestinian terrorists had hijacked a Lufthansa aircraft. Two members of the SAS, Major Alastair Morrison and Sergeant Barry Davies, joined a German assault team subsequently, and rescued the German hostages successfully. The Callaghan government was suitably impressed, and soon after authorised a substantial increase in the CRW force as well as additional funds for improved equipment, training and weapons and communications. A special projects (SP) team was also set up, which involved training its recruits to carry out siege-busting exercises at SAS headquarters in Hereford.

A key part of this was undergoing a close-quarters battle (CQB) course, part of which involved training in the six-room 'Killing House'. Complete with furniture, the Killing House contained paper targets in the form of Russian soldiers representing terrorists, and others representing hostages, which were moved from place to place. The SP team was divided into two specialist groups: the assault group, which stormed the building; and the perimeter containment group, which played the role of snipers and circled the scene, preventing anyone from leaving (or entering), not only at ground level but by way of the sewers or over the roofs.

In Steve Crawford's *The SAS at Close Quarters*, a member of the regiment described the kind of drills conducted around the time of the embassy siege:

> Inside the 'Killing House', live ball ammunition is used all the time, though the walls have a special rubber coating which absorbs the impact of rounds as they hit. Before going into any hostage scene or other scenario, the team always goes through the potential risks they may face. The priority is always to eliminate the immediate threat. If you burst into a room and there are three terrorists – one with a knife, one holding a grenade and one pointing a machine gun – you always shoot the one with the gun, as he or she is the immediate threat.
>
> The aim is to double tap the target until he drops. Only headshots count – in a room that can sometimes be filled with smoke there is no room for mistakes. Hits to the arms, legs and body will be discounted, and constant drills are required to ensure shooting standards are high. If the front man of the team has a

problem with his primary weapon, which is usually a Heckler and Koch MP5 sub-machine gun, he will hold it to his left, drop down on one knee and draw his handgun. The man behind him will then stand over him until the problem with the defective weapon has been rectified. Then the point man will tap his mate's weapon or shout 'close', indicating that he is ready to continue with the assault. Two magazines are usually carried on the weapon, but magnetic clips are used as opposed to tape. Though most of the time only one mag is required, having two together is useful because the additional weight can stop the weapon pulling into the air when firing.

The aim is to slowly polish your skills as a team so that every-one is trained up to the same level, thinking on the same wave length and (being) aware of each other's actions. The 'House' is full of corridors, small rooms and obstacles, and often the scenario demands that the rescue be carried out in darkness (a basic SOP [standard operating procedure] on a live mission is for the power to be cut before the team goes into a building). The rooms are pretty barren, but they can be laid out to resem-ble the size and layout of a potential target, and the hostages will often be mixed in among the gunmen. Confidence in using live ammunition is developed by using 'live' hostages, who are drawn from the teams (the men wear body armour but no helmets). They usually sit at a table or stand on a marked spot, waiting to be 'rescued'. The CQB range also includes electron-ically operated figures that can be controlled by the training staff. At a basic level, for example, three figures will have their backs to you as you enter the room. Suddenly, all three will turn and one will be armed. In that split second, you must make the right assessment and target the correct 'body' – if you don't you will 'kill' a hostage and the gunman will 'kill' you.

A variety of situations can be developed by the instructors. For example, they may tell the team leaders to stand down minutes before a rescue drill starts, forcing the team members to go through on their own. Other 'funnies' include smoke, gas, obstacles to separate team members from their colleagues, as well as loudspeakers to simulate crowd noises and shouting.

Apart from using sub-machine guns and automatic pistols, shotguns – such as the Remington 870 pump-action model – as well as explosives were used to blow off door hinges and locks. They also carried assault ladders, which allowed for the silent scaling of walls, as well as rapid access to buildings, vehicles, ships, aircraft, trains and buses. Once inside, they could then deploy a specially designed stun grenade, the G60 'flash-bang', which was devised to blind and deafen an opponent before the SAS shot him dead.

The SP team was also equipped with specialised equipment that could be used to determine the location of hostages and gunmen inside a building. For instance, fibre-optic equipment could be threaded into a room to view events without the occupants' knowledge. In addition, they could potentially overhear conversations on various types of listening devices and thus, possibly, fix the positions of the hostage-takers. With all of this, the SP team were one of the best hostage-rescue teams in the world, and they would soon be thrown into action.

The SAS had actually learnt of the embassy siege before official notification was even received. At 1144 hours, an ex-D Squadron SAS corporal named Dusty Gray, then serving in the Metropolitan Police as a dog-handler, rang SAS headquarters at Hereford to provide Lieutenant Colonel Mike Rose, commanding officer of 22 SAS, with the little information that he then had. Despite being sceptical, Rose did not wait for official confirmation. Blue Team and Red Team of the SP team were swiftly sent to London and arrived in the early hours of 1 May at Regent's Park barracks.

However, there was initially nothing the SAS team could do other than to sit and wait. The situation was in the hands of the government, the Metropolitan Police and the hostage negotiators, who all endeavoured to find a peaceful solution, without giving in to the terrorists' demands. Meanwhile, as the siege went into its second day, the Blue Team and the Red Team discussed their options for every scenario. Intelligence officers also examined all that they knew about the layout of the embassy building, and where the hostages and terrorists were situated. The embassy caretaker, who had intimate knowledge of the building's layout, was to prove particularly useful and further help was also soon forthcoming.

After substantial deliberations, Oan had no option but to release one of the hostages, who had become ill. This was a huge strategic error. On his release, the hostage was subsequently debriefed about the number of gunmen, their weapons, the layout of the embassy and the location of the hostages. Meanwhile, MI5 had managed to drill holes into the walls of the embassy so that they could install microphones, with the sound deliberately obscured by the use of low-flying aircraft. With all this intelligence, the SAS rapidly built scale models of the embassy, and its rooms, from plywood.

By day three of the siege, as the Blue Team and Red Team were taking it in turns to practise inside the makeshift rooms of the model embassy, things in the real embassy were growing increasingly tense. Oan was now threatening to kill the hostages, demanding to speak to the assembled media and making a series of new demands.

Negotiators continued to buy more time, when on day four two hostages were eventually released in exchange for the BBC broadcasting a statement. Despite this, Oan was beginning to realise that none of his demands would be met. It was clear that the situation could not continue for much longer.

Meanwhile, SAS plans continued apace. At 2300 hours, the team made its way across the rooftops and reached the embassy, where, careful not to make a noise, they proceeded to break the lock on the skylight. Now there was an easy way in, should it be needed. They also secured abseil ropes to several chimneys, so rapid ascent could be made down the rear of the building to the lower floors for entry through the windows. All the while, they continued to practise in the scale model at Regent's Park barracks.

As the siege rumbled into day six, and with no progress being made, Oan grew increasingly suspicious that the embassy had been infiltrated and began to accuse the police of double-crossing him. He told the police negotiator, 'You have run out of time. There will be no more talking.' This was a real concern, particularly when the sound of gunshots was soon heard, followed by a body being dumped outside. It was now or never. Prime Minister Margaret Thatcher subsequently gave the go-ahead for the SAS to proceed.

At 1923 hours, the codeword 'Hyde Park' came over the assault teams' radios. This was the signal for the abseilers on the roof to hitch themselves to their ropes. A few moments later 'London Bridge' – the signal to descend – echoed in Red Team's ears, followed by, 'Go! Go! Go!' Operation Nimrod, the code-name for the SAS rescue, was now operational.

As Red Team 2 descended towards the second-floor balcony, Red Team 1 blew in the glass dome in the stairwell leading to the second floor, from which they entered the building and ran upstairs to clear the third and fourth floors. Meanwhile, Blue Team, working simultaneously, executed an entry through the library at the rear of the ground floor.

However, disaster struck Red Team 2 when its leader got snagged in his harness. As others tried to help free him, someone's foot smashed a window, which alerted Oan. As Oan went to investigate, taking PC Lock with him, Red Team 2 was still dangling to the side of the building, with their plan fast unravelling. Next door to the embassy, Blue Teams 4 and 5 watched on in horror. They were poised to use explosive charges to blow in the rear French doors but this situation made that impossible, as detailed in *The SAS at Close Quarters*:

We took up a position behind a low wall as a demolition call sign ran forward and placed the explosive charge on the embassy French windows. It was then that we saw the abseiler swinging in flames on the first floor [sic: second]. It was all noise, confusion, bursts of sub-machine gun fire. I could hear women screaming. Christ! It's all going wrong, I thought. There's no way we can blow that charge without injuring the abseiler. Instant change of plans. The sledge-man ran forward and lifted the sledge-hammer. One blow, just above the lock, was sufficient to open the door. They say luck shines on the brave. We were certainly lucky. If that door had been bolted or barricaded, we would have had big problems.

'Go. Go. Go. Get in at the rear.' The voice was screaming in my ear. The eight call signs rose to their feet as one and then we were sweeping in through the splintered door. All feelings of doubt and fear had now disappeared. I was blasted. The adrenalin was bursting through my bloodstream. Fearsome! I got a fearsome rush, the best one of my life. I had the heavy body armour on, with high velocity plates front and back. During training, it weighs a ton. Now it felt like a T-shirt. Search and destroy!

However, with Red Team 2's leader still entangled in his rope, and Oan getting ever closer, the team needed to buy time. As such, they threw stun grenades and CS gas canisters into the embassy, rather than enter as had been planned. This only made matters worse. The grenades landed on newspapers, which had been covered in lighter fuel, and started a fire, which began to burn the team leader as he hung suspended in mid-air.

Red Team 2 frantically cut him loose but not before he'd suffered serious burns. While their leader could play no further part in the operation, they finally entered the embassy, which was thick with fire, smoke and gas. Before them, in the smoke, was one of the terrorists. However, as one of the team raised his MP5 to shoot he found it had jammed. The terrorist stared in disbelief. He had thought he was a goner.

At this, he made a run for it, with the soldier giving chase, grabbing his 9mm Browning from the holster strapped to his thigh as he did so. But suddenly the chase became far more urgent. The soldier realised that the terrorist was holding a grenade and was heading directly for a room full of hostages. Taking aim, his vision blurred by smoke, he knew he had just one chance to get this right, otherwise the hostages would all die. The years of training in the 'Killing House' all came to the fore as, with a steady hand, he squeezed the trigger, watching as seconds later the terrorist's head snapped back and his body collapsed to the floor. He had killed him with just seconds to spare. Yet there was to be no respite. Moments later gunshots rang out from the telex room. Three terrorists had begun firing on the hostages, killing one and wounding others.

Red Team immediately rushed for the room. As they entered, a terrorist charged at them with a grenade in his hand. He was instantly shot in the head. This gave the signal to the surviving terrorists that they were outnumbered and outmanoeuvred by the best in the business. Yet they still refused to surrender, instead trying to hide and blend in with the hostages. The soldiers began searching each hostage, when someone made a quick move to their pocket. Without stopping to think, one of Red Team shot him dead. For a moment, they feared they had accidentally killed a nervous hostage. But, when they turned the body over, they found the man had a grenade in his hand. It was a sobering moment.

With the telex room apparently secure, with the terrorists there dead and the hostages safe, the team then cleared room after room, shooting off locks, kicking in doors, and throwing in stun grenades. However, a tremendous explosion suddenly reverberated through the embassy. Red Team was unperturbed. They knew this was just the plastic explosives Blue Team 3 had placed against the window of the rear second and upper floors in order to gain entry.

As one member of Blue Team recalled to Crawford:

> Then we were in. We threw in some grenades and then quickly followed. There was a thundering bang and a blinding flash as the stun grenades went off. Designed to disorientate any hostiles who were in the room, they were a godsend. No one in here, good. I looked around, the stun grenades had set light to the curtains, not so good. No time to stop and put out the fire. Keep moving. We swept the room, then heard shouts coming from another office. We hurried towards the noise, and burst in to see one of the terrorists struggling with the copper who had been on duty when the embassy had been seized: PC Lock.

On seeing the SAS men coming through the back of the building, Lock had rugby-tackled Oan to the ground. The SAS stormed into the office and shouted for Lock to move out of the way. As he did so, they opened fire on Oan, hitting him in the head and chest, killing him instantly.

With all the rooms cleared, and Oan dead, the SAS assembled a line on the main staircase, where they roughly manhandled the hostages down the stairs and out of the back as quickly as possible. It seemed they had succeeded. However, one of the 'hostages' suddenly pulled out a grenade and reached for the pin. Pete Winner, in Gregory Fremont-Barnes's book *Who Dares Wins: The SAS and the Iranian Embassy Siege 1980*, records what happened next:

> He drew level with me. Then I saw it – a Russian fragmentation grenade. I could see the detonator cap protruding from his hand. I moved my hands to the MP5 and slipped the safety-catch to 'automatic'. Through the smoke and gloom I could see call signs at the bottom of the stairs in the hallway. Shit! I can't fire. They are in my line of sight, the bullets will go straight through the terrorist and into my mates. I've got to immobilize the bastard.
>
> Instinctively, I raised the MP5 above my head and in one swift sharp movement brought the stock of the weapon down on the back of his neck. I hit him as hard as I could. His head snapped backward and for one fleeting second I caught sight of his tortured, hate-filled face. He collapsed forward and rolled down the remaining few steps, hitting the carpet in the hallway, a sagging, crumpled heap. The sound of two magazines being emptied into him was deafening. As he twitched and vomited his life away, his hand opened and the grenade rolled out. In that split second, my mind was so crystal clear with adrenalin it zoomed straight in on the grenade pin and lever. I stared at the mechanism for what seemed like an eternity, and what I saw flooded the very core of me with relief and elation. The pin was still located in the lever. It was all over; everything was going to be ok.

It had all taken seventeen minutes. At the start of the siege, there had been twenty-six hostages. Two had been killed by the terrorists while five had been released before the assault, leaving nineteen to be successfully rescued. All were now safe and well, if not a little shaken.

If they weren't well known before the raid, the SAS were now acknowledged as one of the world's finest elite special forces, a moniker they continue to enjoy to this day. Yet the events of 9/11 would soon change the world of anti-terrorism for ever. As the world teetered on the brink of all-out war in the days that followed, the United States looked to invade Afghanistan with just a handful of men. It was a mission that would go down in legend ...

24

THE GREEN BERETS

The world changed for ever on Tuesday, 11 September 2001, when terrorists hijacked four planes, crashing two into the World Trade Center's twin towers and one into the Pentagon. The fourth hijacked plane was believed to be en route to the White House, before passengers thwarted the terrorists and brought it down near Shanksville, Pennsylvania. Almost 3,000 civilians were killed and more than 6,000 injured in the deadliest attack on American soil since Pearl Harbor. Within hours, Osama bin Laden, on behalf of the terrorist organisation al-Qaeda, had claimed responsibility. Incensed, the American people demanded immediate retribution. However, getting to bin Laden was not easy. Intelligence suggested he was being protected by the hard-line Taliban government in Afghanistan, and despite repeated demands by President Bush to give him up, they refused. All eyes now turned to the military.

It was initially suggested that America could invade Afghanistan with 60,000 troops. However, such a massive migration could take six months. Moreover, while there was usually a contingency plan for invading a country, when it came to Afghanistan no such plan existed. Up-to-date intelligence on the country, or the Taliban, was also hard to come by. With all this in mind, it was clear that a traditional military invasion was going to prove very difficult, and costly. Indeed, the US military had already seen the mess the Soviet Union had gotten itself into when invading Afghanistan in the 1980s and they didn't want to make the same mistake.

Nonetheless, a solution soon presented itself. The CIA was at least aware that an Afghan rebel group called the Northern Alliance had been attempting to overthrow the Taliban ever since it had taken power in 1996. This armed force was already on the ground and knew the landscape far better than the US military. If the United States could somehow build alliances with their commanders, through offers of support or bribes, the Afghan armies might be of significant assistance. Not only could the Northern Alliance do much of the fighting, it could also provide crucial intelligence about key Taliban locations, which could then be bombed from the sky with laser precision. Moreover, they might even be able to help find Osama bin Laden.

Indeed, should bin Laden be found, orders were to kill rather than capture. Cofer Black, director of the CIA's Counterterrorism Center (CTC), had made this explicitly clear: 'I don't want bin Laden and his thugs captured. I want them dead ... I want to see photos of their heads on pikes. I want bin Laden's head shipped in a box of dry ice. I want to be able to show bin Laden's head to the President.'

It seemed that the US needed to build relationships with two Northern Alliance warlords in particular, Atta Mohammad Noor, and General Rashid Dostum. However, although Noor and Dostum hated the Taliban, they had often been at war with each other. Furthermore, neither was said to be particularly trustworthy. 'We knew nothing about these guys,' Lieutenant General John F. Mulholland said of this. 'All of these guys have blood on their hands. None of these guys are clean actors.' Earning their trust, and persuading them to join forces with the United States, was not going to be easy.

It was determined that CIA operatives would initially be sent into Afghanistan to meet with Noor and Dostum as the first point of contact. Yet it was soon clear that there was only one force that could be depended on to take on the brunt of this top-secret operation: the United States Army Special Forces, aka the Green Berets.

In 1952, with the Cold War heating up, Colonel Aaron Bank had looked to create an army unit that could operate clandestinely in foreign countries hostile to the United States, drawing particular inspiration from the escapades of the wartime Office of Strategic Services (OSS) and Jedburgh teams. During the Second World War, the OSS was renowned for its sabotage missions against German infrastructure targets, earning the nickname the 'Devil's Brigade', due to its ability to sneak into German positions and slit the enemy's throat. Meanwhile, the Jedburghs were trained to infiltrate occupied countries and train resistance and guerrilla fighters. With this as his inspiration, Bank proceeded to form the 10th Special Forces Group. After adopting their distinctive headgear, this group would be for ever known as the Green Berets.

With the Berets' headquarters at Fort Bragg, North Carolina, candidates were normally expected to have served a number of years in the army, and also to have completed Airborne (parachute) School before being accepted into the Special Forces Assessment and Selection (SFAS) course. This was merely a warm-up course designed to separate the weak from the strong. Right from the start the question was posed: How badly do you want this?

Tests on the twenty-four-day SFAS programme featured timed land navigation exercises (including a 47-mile individual navigation course, with candidates carrying over 100lbs in their packs, to be completed within a 72-hour time period), route marches, obstacle courses and other gruelling physical tests. Situational awareness reaction exercises were also a key part of SFAS, which were designed to test how a candidate could think on his feet, adapt to new situations, relate to others and operate as part of a team.

According to historian Dick Couch, out of the 3,000 soldiers who applied for SFAS each year, only 600 made the grade. Those who did so now passed on to the second stage of selection – the Special Forces Qualification Course. Known as the Q Course, this usually took successful candidates around a year to complete, with them having to pass five fixed phases of training, which included:

1. A seven-week orientation course that gave candidates a thorough understanding of the history, traditions, roles and planning processes of the Green Berets.

2. Learning foreign languages for between eighteen and twenty-five weeks, such as Russian or Arabic.

3. Mastering small units' tactics such as building clearance, close-quarters battle (CBQ), raiding, ambushing, reconnaissance, patrolling and advanced marksmanship. Candidates also undertook SERE (Survival, Evasion, Resistance and Escape) – a three-week training course designed to equip them with the necessary skills to survive in the wilderness and evade capture by enemy forces. This included a five-day resistance-to-interrogation module known as the Resistance Training Laboratory.

4. Military Occupational Specialty (MOS) training, whereby each soldier was trained in one of the core areas – weapons, engineering, medical and communication.

5. Robin Sage: a fully immersive five-week-long field exercise in which prospective Green Berets had to parachute or helicopter into the fictional country of 'Pineland' where they were to find and recruit 'guerrillas' and lead a successful liberation of the country. The task usually involved establishing contact with the leader and explaining how an alliance could benefit the rebels. The guerrilla leader (a role-playing member of the Special Forces) would invariably bluff, bully and threaten the young captain, while trying to extract all the supplies and hardware they could,

ceding as little as possible in return. In effect, it was a test of psychological gamesmanship.

Those who successfully completed the Q Course claimed the prize of their iconic headgear at a formal mess ceremony before being assigned to their Special Forces Group (SFG). Each such group had a particular regional focus; in the case of the 5th Special Forces Group, for instance, it was the Middle East, Persian Gulf, Central Asia and the Horn of Africa. And it was for this reason that, in October 2001, all eyes turned to the 5th to avenge 9/11.

At its Fort Campbell headquarters in Kentucky, Colonel John Mulholland gathered his men together in their team room and delivered the words they had been waiting to hear: 'Gentlemen, you have been selected to infiltrate Afghanistan.' Subsequently split up into teams of twelve, known as Operational Detachments-A (ODAs), each ODA would have two qualified commanders, a captain, a warrant officer and a pair of each specialty: operations and intelligence; weapons; engineer; medic; and communications. If required, this would enable the team to split into two units of six and retain the full complement of skills in each.

ODA 595, led by Mark Nutsch, would be the first team sent in to join forces with General Dostum, while soon after Dean Nosorog would head up ODA 534 to meet with Noor.

Both Nutsch and Nosorog had made tremendous sacrifices to be on the mission. Just days prior to 9/11, 32-year-old Nutsch had been promoted to a staff position at 5th Group headquarters. Wanting to do his duty, and not miss out on the mission of a life-time, he had asked Lieutenant Colonel Bowers to be returned to his team. Nosorog, meanwhile, had actually been on honeymoon when the World Trade Center had been attacked. On seeing the news, he had immediately booked the first flight to Kentucky to be ready to go wherever he was required. Now both men were putting their lives on the line, with the hopes of their country weighing heavily on their shoulders.

On the night of 18 October, Captain Nutsch and his twelve-man team were ferried over the Hindu Kush mountain range, a 16,000ft-high mass of rock and permanent snow, by an MH-47E Chinook helicopter. For now, these twelve men represented the entire American fighting force aiming to take on an entire country, as well as capture the world's most wanted man: Osama bin Laden.

There had already been three previous attempts to get the team into Afghanistan but poor weather, and Taliban resistance, had seen them all aborted. The chopper now flew with all its lights off, and without heating, so as to avoid the attention of any heat-seeking missiles. The heavy snow and winds made the journey through the dark mountains extremely hazardous.

Already bitterly cold, Nutsch and his team were well aware they were likely to die on this operation; indeed, many were scared they'd be shot out of the sky by Taliban anti-aircraft missiles before they had even set foot in Afghanistan. For Nutsch, the best-case scenario was taking down the Taliban over the course of the next year or two. He was certain this wouldn't be a quick affair. Striking up a relationship with Dostum, and getting intelligence from him, would be tough enough, while the cold Afghan winter was fast approaching, which would make large parts of the barren country inhospitable until the snow thawed in spring. Nutsch was therefore prepared for the long haul.

However, while their job was to build a relationship with General Dostum and his army, some of the team were apprehensive. Some spoke Russian, and a sprinkling of Arabic, but such had been the dearth of intelligence available on Afghanistan, and Dostum, that many of the team felt unprepared. To fill in the gaps, they had devoured any books or magazine articles that were available on the country. A favourite was *The Bear Went Over the*

Mountain, which detailed the Soviet experience in Afghanistan, something they would have to learn lessons from if they were to have any success this time around.

In the mid-1980s, I actually came very close to flying to Afghanistan to help them fight the Soviet invaders. I saw it that my grandfather had fought for the empire, my dad had fought the Nazis, and I wanted to fight the Marxist threat. In Germany with the Scots Greys, I had spent years waiting for the armed might of the Soviet Union to attack, but this never materialised. So, for a chance of real action against the Soviets, I fancied joining the mujahideen guerrillas in their armed jihad. However, such an ambition was soon to be thwarted. Firstly, some research led me to reading that the Afghan authorities weren't much better than their Soviet invaders. According to Amnesty International, 27,000 Afghan civilians had been executed by the Kabul government before the Soviet invasion, and after it hundreds of thousands had been killed in bombing raids or massacres. I certainly didn't want to have that uneasy feeling I once had in Dhofar, whereby I wasn't sure I was fighting for the right side. Thankfully, a job offer to act as public relations officer for Occidental Petroleum in the UK saw me forget about jihad in the mountains and instead prepare for the nine-to-five London rat race. In hindsight, that is definitely one of the better decisions I have made.

The Berets were aware that fitting in would play a crucial part in being accepted by the Afghans, so, as part of their preparations, Nutsch and most of his team had begun to grow to beards. In Afghan culture, facial hair was considered a necessary rite of passage for adult males, and any man without a beard was not taken seriously. They also carried with them black and white cotton keffiyehs, which would not only keep them warm when wrapped around their necks and faces, but were also worn by many in the Northern Alliance.

To help get off to a good start with Dostum, they came bearing gifts: oats for his horses, medical supplies and blankets for his men, and bottles of vodka for Dostum himself. They were aware that Dostum was not your average Muslim warlord. For one, he loved booze, and he was also said to have a liking for hookers. Nutsch knew he would have to bide his time to get to know him, while he also remembered Colonel Mulholland's warning: 'Trust no one!' For that reason, they would have to rely on their training, and their wits, as well as their array of weapons, which included M-4 carbine rifles, grenades, and a 9mm pistol.

In the dark, the chopper finally located its landing zone in the Darya Suf river valley. Disembarking into the cold desert night, Nutsch and his team found Mike Spann, J. J. Sawyer and Dave Olson of the CIA, as well as some of Dostum's men, waiting to meet them. Carrying their rucksacks, they followed the men to Dostum's camp just a few hundred yards away, where they would rest for the night before meeting the general in the morning.

However, despite the friendly welcome, the men, dressed in their US Army tan fatigues, kept their hands close to their 9mm pistols holstered to their right legs. While they had been warned to trust no one, they were also aware that the Taliban had put a bounty of $100,000 on each of their heads. In a poor country like Afghanistan, this was an absolute fortune. They would need to exhibit extreme caution at all times. For this reason, even their names and insignia had been removed from their uniforms, not only for their own safety, but also so that no one could track them back to their families in the US and do them harm.

At dawn, Nutsch woke to find that Dostum and his men had arrived from their mountain hideout. Doug Stanton, in his tremendous book *12 Strong*, which details the Green Berets' time in Afghanistan, recalls how Nutsch was sure to obey local customs upon meeting the 6ft-tall, gregarious Dostum for the first time.

'He put his right hand over his heart and said, "*Salaam alaikum.* Peace be unto you brother."'

Sitting on a carpet, as Dostum's men provided local delicacies such as pistachios, almonds and apricots, Nutsch proceeded to offer the general the gifts they'd brought with them. While they were gratefully received, particularly the vodka, it was clear that Dostum would have really preferred bombs. But, Nutsch had something far better to offer: laser-guided bombs dropped by the US Air Force with incredible precision. Dostum was eager to see these in action – immediately.

In order to do this, Nutsch would require the exact co-ordinates of the Taliban's position. This would involve travelling to Dostum's mountain hideout on horseback, something hardly any of the men had been trained to do. But, within minutes of Dostum's proposal, half of Nutsch's team were clambering onto ragged horses and, not wanting to show the Afghans they were apprehensive in any way, they set off on the arduous journey.

Learning to ride as they went along, some of the men found their horses to have minds of their own and were almost sent hurtling off the sides of the mountain. Others were unused to sitting in saddles and soon developed painful blisters that were red raw. With a little bit of luck and persistence, most of them soon got the hang of how to handle the animals, which was just as well since, for the coming weeks, they would be the Special Forces' primary means of transport.

In an interview with PBS, Captain Will Summers recalled the experience of learning to ride in such circumstances:

And my horse turned and faced straight down the hill ... And he crouched down like a cat, and just sprung off the side of the mountain. And, I think about three to five horse lengths later, his front feet hit. And this guy just took off like lightning down the side of a cliff. The only thing that went through my mind was this 1980s movie, *The Man from Snowy River*. And so, I was like, 'Okay, the guy from Snowy River, he put his head on the back of the horse, and he put his feet up around his neck.'

And so, my feet came up, my head goes back. And I have like horsetail on the back of my head. And this guy just tears down the side of this mountain where at the bottom of it is like a gully about 6–12ft deep, and about 4ft wide ... And he successfully jumped over that ...

And I guess about twenty minutes later, the general [Dostum] and some of his entourage had finally caught up. And he had stopped, and looked at me kind of strange again, but a little different this time. And, he said something to me. And he started off again on his horse. And he turned around, and he said something again. And I knew that he was pretty serious about what he was saying. And then we walked off. And his translator said, 'The general just paid you a great compliment.' And I was like, 'Wow, that's great. What did he say?' And he said, 'Truly, you are the finest horseman he has ever seen.' And then he had stopped and said, in addition to this, I was the most daring and brave man he had ever known.

While the Special Forces had to embrace an old-fashioned mode of transport, they found that the men in the Northern Alliance, who travelled with them, also looked like a medieval army. Bedraggled and bearded, the men carried battered weapons that had either been seized from the Soviets in the 1980s or had been cast-offs from other long-ago wars. For instance, bayonets that had the year '1913' stamped on them were still found to be in use. Still, these men were eager to fight the Taliban and earn their freedom. For the Americans, this made them essential.

Passing through the Darya Suf valley, they saw many empty villages that had been wiped out by the Taliban. The buildings had been burnt to the ground, the women raped and murdered, and the men beheaded. It was a bitter reminder that, while the Special Forces had their own reasons for being in Afghanistan, they were also helping its people take back their country from fanatics.

Upon reaching the foot of the mountain, after many hours of travelling, Nutsch and his team now faced another terrifying prospect. Ascending hundreds of feet above the ground, on just a narrow path with no barriers, they had to hang onto their wild horses for their lives. For men who had virtually no horse-riding experience, this was a hair-raising prospect. Yet everyone stayed on their horse and soon they had reached Dostum's hideout, from which the whole valley stretched below. After twenty-four hours of little sleep, and an exhausting journey, Nutsch now had to prove his worth. If he failed, Dostum might decide that he could do without the Americans after all.

Looking out over the vast valley, Nutsch asked Dostum to point out the Taliban positions. When Dostum did so, Nutsch could not be sure if the Taliban were in fact situated where he was pointing. He was under strict orders to avoid bombing civilians at all costs, not least because this would only give the Taliban propaganda to use against them. To reassure Nutsch, Dostum contacted the Taliban via radio, where in the course of the conversation he not only proceeded to insult them mercilessly, but also managed to get them to confirm their positions.

With this, Nutsch was satisfied and he used his GPS, and a map, to work out the position's latitude and longitude before radioing a B52 overhead. He handed Dostum a pair of binoculars and the warlord watched on as seconds later a 1,200lb bomb dropped through the air and wiped out the surrounding area. Dostum was seriously impressed. It would have taken weeks, if not months, to attack the Taliban in their mountain forts, losing many men in the process. Now, in just a matter of seconds, they had been destroyed. Clearly Nutsch and his team were going to be of some use.

That night Dostum invited the Americans to eat with him. Sitting on a rug, the team enjoyed chicken and lamb, fresh salad, rice and flatbread, remembering to scoop the food up with their right hand, as the left hand was considered unclean in Islamic cultures. This was a habit I also picked up in Dhofar, so much so that, while I remain right-handed, when it comes to issues in the lavatory, I am left-handed, just for this very reason. I also remember ensuring I was barefoot when eating, with my legs tucked underneath so that the soles of my feet could not point at another man – for this would be an insult to a Muslim. Sadly, I did not always manage to avoid this. When I first ate with my men, they kindly picked out the most succulent pieces of meat from their side of the communal dish and gave them to me. I assumed it would be polite to return them but I soon learnt this would cause grave offence. As such, I was happy to eat the succulent meat, while also ensuring not to insult anyone. This did not, however, always work in my favour. When on patrol, villagers would often urge us to share their food with them, with sheep's eyes being their main delicacy. I quickly learnt to swallow while smiling. However, while my Arabic was initially poor, I did acquire a phrase I could use in any number of situations – *Insha'Allah*, which means 'God willing'. As this was non-committal, I soon came to say it whenever I was asked a question and wasn't sure of the answer.

Unlike myself, Nutsch and his team were thankfully experts at ingratiating themselves with their hosts. With trust established, Dostum revealed to Nutsch the city they should target if they were to take down the Taliban: Mazar-i-Sharif.

While rich oil and gas deposits lay near the city's airport, it also boasted the country's longest paved runway, capable of landing transport and supply aircraft. The bridge over the Amu Darya River could also be used to move men and material from Uzbekistan. If the Northern Alliance could take Mazar, then Kabul, the capital, would fall soon after, and then all of Afghanistan with it.

With this target in mind, Dostum and ODA 595 spent the coming days nailing down Taliban targets and unleashing bombs on their positions. As one Green Beret told David Tucker and Christopher Lamb for their book *United States Special Operations Forces*: 'The hard part is developing the infrastructure that facilitates knowing where the targets are, so you can bomb them. And that is what won the war. It's getting the targets and getting which targets, why this target or why that target, and that's what makes it work.'

The Berets didn't have the manpower or the weapons to over-throw the Taliban by themselves. But what they did have was the training to build trust and to extract information that was crucial to crippling the Taliban regime. One minute the team would be on horseback, the next they would be using cutting-edge technology to guide bombs from 20,000ft.

Soon they had further help, when another team arrived with a special operations laser marker (SOFLAM). This made targeting a Taliban tank or truck far easier than the GPS method Nutsch had been using. Once the laser was locked onto a target, the bomb could hit it with unerring accuracy. It could even follow vehicles on the move. As long as the laser remained on the target, the bomb would track it.

However, Nutsch realised that even though they were obliterating Taliban positions they weren't doing as much damage as they could. He needed to lure the Taliban out in large numbers, along with their tanks and trucks. This would require Dostum and his men facing them in battle, and preparing to charge. When the Taliban then assembled, Nutsch could order in the bombs. Following this, Dostum's men would finish off any survivors as they lay wounded or tried to run. In theory, this was a brilliant move. In practice, it almost ended in disaster.

In the heat of battle, orders became confused, with Dostum telling his men to charge just as Nutsch had ordered a bomb to drop. Nutsch could only watch in horror as the Afghans thundered towards the Taliban, praying that the bomb would hit before they reached the position. If it were to wipe out hundreds of Dostum's men, Nutsch knew it would be hard to recover the relationship.

As the Taliban unleashed their guns on the Afghans, and their tanks prepared to open fire, the earth suddenly shook with a massive explosion. Nutsch looked through the dust for any sign of Dostum's men. To his relief, with the dust slowly dispersing, he saw them, still alive, shooting, stabbing and even beheading any survivors.

Yet, as they took village after village, Nutsch and his team didn't always have the luxury of fighting at arm's length. With one particularly vicious fight going against the Northern Alliance, Dostum entered the field to inspire his men. Nutsch and his team knew they had to join him. To lose Dostum would be a disaster, as it would leave them in the hands of his successor, who might have very different ideas about an alliance with the Americans. On horseback, they subsequently roared into battle, firing their rifles at the oncoming Taliban, driving them back and ensuring Dostum was kept alive. It was like a scene from the First World War, certainly one the Royal Scots Greys would have been proud of.

Moving ever closer to Mazar, the Berets soon came across a deadly obstacle that had haunted Afghanistan for decades: a minefield. It was only by the grace of God that no one was killed. But, stuck in the middle and without the use of mine detectors, it appeared they were trapped. The Americans could only watch in awe as the Afghans subsequently bent down on their hands and knees and painstakingly cleared the mines by softly lifting them from their hiding places and disarming them.

During my time in Dhofar, the *Adoo* had covered the ground upon which we regularly travelled with anti-tank mines. In the supposed safety of a Land Rover, I don't recall feeling that concerned, but this was in spite of the evidence I saw every day, as the mines caused some horrific injuries, sheering off limbs and killing many men.

Rather than an actual minefield filling me with dread, during an unassisted attempt to cross Antarctica, I experienced something similar that truly scared me. Mike Stroud and I were walking across the snow-packed ice when suddenly a gaping hole opened up just ten paces ahead of Mike. It was 45ft wide and 120ft in length. It was a miracle it didn't swallow him whole. But we still weren't out of trouble. All around us, renewed explosions announced further cratering. It was terrifying. We couldn't stay where we were, as the ground was now unstable, but we also didn't know where the next crater would emerge, or for how long this could continue. Just like a minefield, we had to just keep going, never knowing when the ice shelf would rumble and reverberate around us again. It was only sheer luck that we weren't plunged into the icy waters below.

As Nutsch and his team continued with Dostum's army towards Mazar, it was now time for ODA 534 to meet up with Afghanistan's other notorious warlord, Atta Mohammad Noor. It was not only important that Captain Nosorog establish a relationship with Noor, but he also needed to persuade him to fight on the same side as his rival. After all, Dostum and Noor had a common enemy they hated more than each other – the Taliban.

On 2 November, Nosorog finally met with the bearded Noor at his camp near the village of Ak Kupruk. Noor was delighted the Americans had arrived. He had been fighting the Taliban for five years yet no one in the west had listened to his warnings or cries for help. Now his efforts to resist would finally be rewarded.

Over the years, he had built an impressive list of informants. Some lived in Taliban towns while others served in the Taliban itself. With this intelligence, ODA 534 embarked on a similar journey to that undertaken by Dostum and Nutsch, finding Taliban positions and bombing them into oblivion, all the while moving towards Mazar, where hopefully their two armies would join together and rout the Taliban.

With two Special Forces teams in Afghanistan, working with two warlords, Lieutenant Colonel Max Bowers arrived at Dostum's camp to oversee the operation. After all, it would be disastrous should either side accidentally bomb the other in their haste to take out the enemy. Moreover, there was still concern that the warlords could fabricate positions to take out the forces of their rival. Thankfully, it didn't come to that.

Yet, while taking down the Taliban proceeded at a speed no one had thought possible, the whereabouts of the Special Forces' chief target, Osama bin Laden, remained a mystery. With a huge bounty on his head, there was certainly no shortage of leads. Sadly, these often turned out to be just wild guesses, with Afghans eager to earn the reward. However, when the Taliban government announced on 8 November that it had granted bin Laden Afghan citizenship, it at least seemed that he was still in the country. Some good intelligence even suggested he had been sighted outside Mazar. Such information drove the teams on with a renewed energy to take the city.

When Mazar did eventually fall on 10 November, it all seemed a little too easy. The Americans' intense bombing campaign, coupled with the Northern Alliance sweeping all before them, saw the Taliban flee the city before any great battle could take place. Nosorog and Noor's army were to be the first on the scene and were greeted as conquering heroes by crowds of grateful Afghans. However, while most of the Taliban had left the city, hard-core al-Qaeda members, vowing to die fighting rather than surrender, had holed up in an abandoned school.

In order to avoid more bloodshed, Noor sent some of his men into the school to try to persuade the al-Qaeda fighters to surrender. Soon after, Dostum and Nutsch arrived in the city and were informed of the siege, and that Noor's men were inside. They were told not to attack, or order any bombs to strike the building, before Noor's men were safely out. But Lieutenant Colonel Bowers was not made aware of this conversation. With communication breaking down, a strike was subsequently ordered on the school. All the al-Qaeda operatives were killed, along with Noor's men. Unsurprisingly, this did not go down well. Nosorog had to use all his powers of diplomacy to keep Noor onside and prevent the alliance collapsing just as they had taken their primary goal.

Despite this mishap, al-Qaeda looked to be finished in Mazar and soon the Americans found their huge stash of weapons, including rifles, rockets, grenades and mortars, hidden in a pink building many believed used to be a school. Al-Qaeda therefore had very little left to fight with.

With Mazar a Taliban-free zone, reports came thick and fast of other cities falling, including the capital, Kabul. As the American teams established a base in the city, they now turned their attention to taking the city of Kunduz, which was turning into the place where the Taliban, and al-Qaeda, were making their final stand. While Kunduz was bombed relentlessly, and all roads in and out were cut off, Bowers prepared to move his ODAs in to oversee operations.

However, in the early hours of 24 November, over 600 Taliban fighters suddenly turned up in Mazar and offered to surrender. This was totally unexpected. What's more, no one was sure if they were genuine. A decision was finally made to transport them to the city airport, where they would be held, in lieu of anywhere more suitable.

Due to Afghan custom, none of the men who surrendered were searched. To search a man in such circumstances was considered a gross insult, no matter the circumstances. Despite concerns, the Taliban fighters seemed to be volunteering their weapons before they were taken away. Nosorog was very uneasy with this but didn't want to rock the boat. Besides, he would soon be joining up with Nutsch and moving out to Kunduz.

Yet, as Nosorog and Nutsch left the city with their teams, Dostum suddenly redirected the prisoners to be held at the Qala-i-Jangi fortress, whose walls could hold them more securely than the airport. But Dostum was not aware that within the fortress was the pink school building, where al-Qaeda's weapons stash had been found, and where they still remained.

Stationed at the fort were Johnny 'Mike' Spann and Dave Olson of the CIA. As the prisoners arrived, the CIA men were unaware that they had not been searched but instantly knew this was a bad situation. If they should break out, they could storm the weapons supply in the schoolhouse and avail themselves of enough arms to kit out their 600 men. Moreover, the 600 prisoners significantly outnumbered the remaining Afghan and American forces in the city, which at the time came to only around 160 men, just eight of whom were US Special Forces. They would be able to retake Mazar with ease and then attack the Americans, who had been focusing on Kunduz, from behind. Soon everyone's worst nightmares were realised.

Upon screaming 'Allahu Akbar', the hordes of prisoners set off grenades they had hidden in their clothing and stormed their way out of the basement. In attempting to stop them reaching the school building, Mike Spann was viciously murdered, becoming the first US casualty of the war. Olson managed to escape from them but could not stop the men getting to the weapons stash. A full-blown crisis was now unfolding. If the Taliban escaped from the fort with their weapons, the war could suffer an almighty setback.

With events spiralling out of control, Major Mark Mitchell was unable to call in an airstrike because Olson was unaccounted for but believed to still be alive and inside the fort. As he desperately tried to reach Olson, Northern Alliance soldiers positioned themselves around the walls and opened fire. In response, the Taliban set up mortars in the courtyard and attempted to blow themselves out, taking the Northern Alliance with them. Time was now of the essence.

Frantically radioing Olson, Mitchell finally made contact, only to find he was hiding inside the fort and had no way out. There was also no way Mitchell could rescue him against 600 heavily armed men. There was only one option available. Mitchell called in an airstrike, setting the bombs slightly away from Olson's position. It was risky, and could kill Olson in the process, but time was running out and it was the only play they had. Mitchell told Olson to use the resulting explosion as a diversion and make a break for it, as long as he wasn't killed by the bomb itself.

As the building suddenly rocked with multiple explosions, Olson fled through the debris, using the dust as cover, desperately searching for an exit. Meanwhile, as more bombs continued to drop, the Taliban fled to the basement with their weapons, intending to see out the siege. Many had now been killed but the Northern Alliance had also suffered significant casualties. Luck would, however, soon be on their side.

The US bombers had run low on fuel and had to return to base. Without the US bombers, the Taliban suddenly had the upper hand, even if they did not know it yet.

Attempts to destroy the basement, or force the Taliban out, all failed until the Northern Alliance had a stroke of inspiration. Diverting a small stream to a hole in the basement roof, they proceeded to flood it, giving the Taliban a stark choice: drown or surrender. Soon the remaining eighty-six Taliban survivors gave themselves up, including the American convert John Walker Lindh. The Battle of Qala-i-Jangi was over, as was the threat to Mazar, which was now firmly in Northern Alliance and American hands.

From this point, more Special Forces ODAs entered the country and by December the Taliban had fallen. The first phase of the Afghan campaign was over, and all in a remarkable forty-nine days of combat operations. As Linda Robinson points out in her modern history of the Special Forces, *Masters of Chaos,* 'No one had ever imagined that fewer than 100 Special Forces soldiers and an indigenous militia could overthrow a government so quickly.'

It was a staggering achievement. No other entity within the US military could have done what the Green Berets managed in such a short space of time.

However, the primary goal – to kill Osama bin Laden – had still not been accomplished. The latest intelligence had placed bin Laden, and other key al-Qaeda operatives, in the Tora Bora cave complex. Unmanned aerial vehicles, known as 'Predators', swept the mountain range for any clue as to his whereabouts, while jets dropped bomb after bomb hoping to kill him. Meanwhile, teams of Special Forces searched well over 200 caves. While they found some weapons and previous hideouts, the man they were looking for was nowhere to be found. It seemed he had managed to get over the border and into Pakistan, and all but disappeared.

When he was finally found, after a decade of searching, it would require the unique skills of another elite US military team for the United States to finally have its revenge ...

25

US NAVY SEALS

AD 2011

The nine-year anniversary of 9/11 was fast approaching and still Osama bin Laden was nowhere to be found. The United States government had thrown huge resources at trying to locate him but they had drawn a blank. It seemed not even a $25 million bounty on his head could inspire anyone to give him up. However, interrogating al-Qaeda detainees in Guantanamo Bay had at least provided a lead – the identity of one of bin Laden's most trusted couriers, Abu Ahmed al-Kuwaiti. Armed with his mobile phone number the CIA was able to track his movements, which led them to a suspicious-looking three-storey compound in the suburbs of the Pakistani city of Abbottabad.

Soon the CIA was using everything at its disposal to confirm whether this was indeed where the most wanted man in the world was hiding. The clues were compelling. Satellite images showed that the compound had been built and fortified after 9/11. Its extensive security features included high walls topped with barbed wire, security gates, as well as a privacy screen on all balconies, which prevented anyone from looking inside. It was clearly custom-built to hide someone of significance.

CIA spies proceeded to rent a home that overlooked the compound and they uncovered further suspicious behaviour. The compound had no phone or internet connection, with inhabitants driving for ninety minutes before putting a battery in their mobile phones to make any calls. All rubbish was also burnt inside the high walls, rather than being put out for collection, which made it impossible for the CIA agents to test for DNA. Everything seemed to match but there was still no positive ID of bin Laden. However, a CIA review of all the intelligence available concluded to a very high degree of probability that he was in the compound. Yet getting to bin Laden would not be easy.

President Obama certainly did not want to ask the Pakistani authorities to apprehend him. He was not sure they could be trusted and didn't want to burn what seemed such a strong lead. As a senior adviser to the president later told the *New Yorker* magazine, 'There was a real lack of confidence that the Pakistanis could keep this secret for more than a nanosecond.' A military invasion was also out of the question, with Pakistan supposedly an ally of the United States. Besides, any large-scale military action would only alert bin Laden they were coming for him. As such, Obama realised this was an operation that was perfect for a small elite force to cross into Pakistan, storm the compound and take bin Laden before anyone even knew what was happening. It was just the sort of mission for which the Navy SEALs were made.

An acronym that stands for Sea, Air and Land, the SEALs can trace their origins back to the Second World War, after a lack of intelligence, and preparation, saw the Marines massacred by the Japanese as they tried to get ashore at Tarawa. From this disaster, it was clear that a professional navy combat demolition unit needed to be developed. This was subsequently set up at Fort Pierce, Florida, before the SEALs were officially born on 1 January 1962, thanks to the enthusiasm of President Kennedy for their guerrilla warfare.

Not just any soldier could become a SEAL. The assessment was one of the toughest, most challenging and brutal experiences anybody could have. Like the Berets, prospective candidates had to pass a variety of tests before they were even allowed to apply for the selection and training programme (known as BUD/S). Applicants had to be twenty-eight or under, have excellent vision, come from certain navy ratings, have the endorsement of their commanders, and have plenty of time remaining on their enlistment before they could even try out. After that, they still had to pass a very strict physical examination at Naval Special Warfare Preparatory School in Great Lakes Illinois. Most recruits are out of the reckoning before they even get a chance to try the twenty-six-week BUD/S selection programme.

Conducted at Coronado, California, right on the Pacific Ocean near San Diego, the BUD/S quickly weeds out those who don't belong in the company of the elite. Only the strongest survive and sometimes not a single recruit makes the grade. Perhaps the most famous part of the BUD/S is what is known as 'Hell Week'. Just before midnight, instructors wake up recruits with incessant screaming and firing off blanks from an M60 machine gun. So begins five and a half days of constant activity that includes running, boat drills and plenty of swimming. All the while, the instructors shout, 'The only easy day was yesterday!' A prospective SEAL sleeps at most four hours during the entire week, runs more than 200 miles, and does physical training for more than twenty hours per day.

A SEAL officer explained why Hell Week is necessary to author Hans Halberstadt for his book *US Navy Seals*:

I continually tell my troops while they are training, 'Look, I can't give you the feeling of what it is really like to be in combat ... because I can't shoot at you and make you hurt. It's illegal, and I wouldn't want to do that anyway. What I can do is to make the conditions so tough, and try to make you so tired, put you under such stress, that you will get something of a feeling of what it is like.

Under the most intense conditions, recruits have to take part in simulations on land involving live ammunition and explosives. They must also attend 'jump school', at Fort Benning, Georgia, where they train to become fully qualified military parachutists. And those who are still standing at the end of all this must then undertake the SEAL Qualification Training (SQT) course that provides students with the core tactical knowledge they will need to join a SEAL platoon.

With President Obama looking to the SEALs to storm bin Laden's Abbottabad compound, he gave Admiral William McRaven three weeks in which to devise a plan. McRaven knew this was a job for his best of the best. As such, he handpicked a team, drawing from the most experienced and senior operators from Red Squadron, of SEAL Team Six, the most elite counter-terrorism unit administered by US Naval Special Warfare Command. These men were perfect for this mission. Their unique training had seen them specialise in storming buildings and killing enemy fighters inside, while the team also had language skills and experience with cross-border operations into Pakistan.

In his book *The Operator*, Robert O'Neill recalls his experience in trying to join this elite SEALs team:

Guys have to complete a vastly amped up version of the physical test that conventional SEALs have to pass. Every distance is longer, every time faster and every exercise has more reps. It's a pretty serious test. If you pass that, then you go before a board of senior officers and enlisted in a room where they grill you for an hour about your array of service medals, your tactics, your experiences, your bosses, your home life, and how much you drink.

After recently returning from Afghanistan, the team were told to report to Harvey Point Defense Testing Activity facility in North Carolina. No one had any idea what was to be discussed but some believed that, with the situation worsening in Libya, it had something to do with taking out Colonel Gaddafi. The chosen few were instead stunned to hear from their commander, 'We think we have found Osama bin Laden, and your job is to kill him.'

As all known intelligence was shared with the team, a plan of action began to take shape. A life-size replica of the Abbottabad compound was subsequently constructed, using CONEX shipping containers. All the measurements were exact, or close to it, and the training was unusually realistic, with the SEALs brought out in conventional Black Hawks and fast-roping out of the choppers into the compound, just as they would for the real thing.

Training for hours and hours with the sun up, they also prac- tised at night so as to get used to doing it wearing night-vision goggles. However, some were growing increasingly wary that the compound might have heavily armed guards, as well as be rigged with explosives. Indeed, that was if they managed to get to the compound, as there was a real fear that Pakistani air defence might shoot the choppers down, with them having no authority to be in the country. Thankfully, the brand-new state-of-the-art helicopters that would be used in the raid were designed to be quiet and to have low radar visibility to avoid this fate.

Every factor seemed to be covered. Nothing was left to chance. Such was the attention to detail that the helicopters were even tested in Nevada, so that the pilots could get used to flying them at a similar altitude to that in Abbottabad.

McRaven presented the planned assault to President Obama and the mission to storm the compound was given the go-ahead on 29 April. Seal Team Six now prepared itself for the raid of a lifetime: Operation Neptune Spear.

Going in light was vital if the SEALs were to take the compound by stealth. Robert O'Neill details in his book how he pared his kit down to the bare minimum. No longer carrying a knife or a pistol, all he had with him was ceramic body armour, a Nalgene water bottle, two protein bars, and a Heckler & Koch 416 automatic rifle, with three extra magazines, along with PVS-15 night-vision goggles.

With their training complete and equipment ready, on the moonless night of 1 May 2011, the twenty-six SEALs boarded two modified Black Hawk helicopters at Bagram airbase in Afghanistan and crossed over the border into Pakistan. Waiting on the outskirts, should they be required, were more SEALs in Chinook helicopters. Tension was high. It would take ninety minutes to reach the compound and, with the Pakistanis unaware of the mission, they could shoot the Black Hawks down as hostile aircraft at any moment. Some of the SEALs chose to sleep, while others focused on the mission ahead, remembering the horrific events of 9/11 as extra motivation. O'Neill remembers being excited at having the opportunity to take down the man who had encouraged countless atrocities around the world, but he was also prepared for the worst-case scenario – that bin Laden might not even be at the compound.

Two minutes away from their target, the doors suddenly opened. With the city lights of Abbottabad twinkling below, moments later the compound came into view, just as it had looked in the training facility. This gave the men extra confidence. Without ever having set foot in the place, they knew every square inch.

The two Black Hawk helicopters now separated and made for their landing zones. But, as Dash 1 hovered over the compound for the SEALs to fast-rope down onto the roof, it suddenly lost control. Entering a 'vortex' condition, due to a hotter than expected air temperature combined with the compound's 18ft-high walls, the pilot had no option but to enact an emergency crash-landing. As he did so, the tail and rotor smashed into the walls, damaging the helicopter beyond repair. No one was hurt but those inside the compound now knew they were under attack. All the SEALs' careful plans were now out of the window. With no time to waste, they quickly disembarked, opening fire as they charged towards the ground floor of the concrete building.

On seeing the crash-landing, the pilot of Dash 2 knew that he couldn't risk entering the compound in the same conditions. As such, he landed just outside the compound walls. The helicopter was safe but the SEALs now had to blow open a gate with explosives, losing vital seconds. Finally inside, they headed for the three-storey building only to come under fire from behind the door of a guardhouse. While still on the move, one SEAL dropped the shooter dead with a single shot.

Blasting their way inside the building, the two SEAL teams heard the cries and screams of women and children before the power suddenly went out. With potential enemies lurking in every corner, the SEALs quickly deployed their night-vision goggles. Going from room to room, with their guns raised, they herded all the women and children together, restraining them with plastic zip ties, while others continued to scale the stairs of the dark compound. So far there was no sign of their target, but CIA intelligence indicated that, if bin Laden was in the compound, he would be found on the third floor.

However, further seconds were lost because the stairways were barricaded shut, requiring more explosives before the SEALs could move upwards. All these hold-ups were allowing whoever was on the next floor to prepare their welcome, and the SEALs were not to be disappointed.

As they made their way to the first floor in total darkness, a figure suddenly darted behind a bannister, armed with an AK-47. Intelligence had suggested this could be Khalid, bin Laden's son. A brief standoff ensued before a SEAL uttered the words, 'Khalid, come here.' Confused by this, Khalid glimpsed around the bannister only for the SEAL to shoot him in the head. A quick inspection confirmed that the dead man was indeed Khalid bin Laden. This only increased confidence that his father wouldn't be too far away.

Progressing quickly to the third floor, clearing all the rooms, the team found two women hiding behind curtains. This was no time for niceties. For all the SEALs knew, they could have been wearing suicide vests. Acting quickly, the point man dragged them to the floor, intending to take the blast himself. Thankfully, they were unarmed. Robert O'Neill, now turned to a room and saw Osama bin Laden before him. He describes the moment in his book:

> Osama bin Laden stood near the entrance at the foot of the bed, taller and thinner than I'd expected, his beard shorter and hair whiter. But it was the guy whose face I'd seen 10,000, 100,000 times. He had a woman in front of him, his hands on her shoulders. In less than a second, I aimed above the woman's right shoulder and pulled the trigger twice. Bin Laden's head split open, and he dropped. I put another bullet in his head. Insurance.

Word soon reached Admiral McRaven via radio: 'For God and country, Geronimo, Geronimo, Geronimo, EKIA [Enemy Killed in Action].' Following the crash-landing, the mission had been accomplished in just fifteen minutes. Watching a live stream of the raid in the White House Situation Room, President Obama smiled, 'We got him.' But the SEALs still needed to get out alive.

Before they left the compound, the SEALs quickly tried to grab as much evidence as possible, with computer hard drives particularly important. As they seized what they could, they also found a huge amount of opium and a sprinkling of pornographic magazines. While they left some of this behind, they made sure to take with them the most important evidence of all – bin Laden's body.

They now had to move quickly. A neighbour had witnessed the raid and was proceeding to upload the latest developments to his Twitter account. Surely the Pakistani military would soon be on the case. The last thing the SEALs needed now was to engage with supposed allies.

Having taking photos of bin Laden's body, the SEALs put him in a body bag and carried it outside. Yet, before they could leave, they had to blow up the Black Hawk that had crash-landed, so as not to allow such state-of-the-art technology to fall into the hands of the Pakistani military.

Bounding inside a waiting Chinook helicopter, they were soon rising above Abbottabad and heading back to Afghanistan. Despite the success of the mission, it was another tense ninety minutes on the return flight to Bagram. The Pakistani military was by now well aware of the operation within its borders and the Chinook was not a stealth bird. It could be tracked on all radars and wouldn't be difficult to intercept or shoot down. Thankfully, the team made it back to base without further incident.

Once bin Laden's body had been inspected by experts, McRaven confirmed to President Obama that the mission had been accomplished, with no SEAL casualties. In front of the world's media, President Obama then said the words the nation had been waiting ten long years to hear:

> Tonight, I can report to the American people and to the world that the United States has conducted an operation that killed Osama bin Laden, the leader of al-Qaeda, and a terrorist who's responsible for the murder of thousands of innocent men, women and children.

For the United States, bin Laden's death finally symbolised some form of justice and retribution. And if SEAL Team Six had been shrouded in secrecy before the raid, it was now on the lips of everyone in the world, ensuring it would become the standard-bearer of elite military units for the current generation.

26

THE FUTURE

Although elite special forces continue to shape the world in which we live their roles have somewhat altered in recent years. There are of course situations that will always require specially trained soldiers putting their boots on the ground. However, nowadays, situations that once required such interventions seem increasingly to be dealt with at arm's length.

Since the invasion of Iraq in 2003, there has been a considerable rise in so-called 'professional security companies' (PSCs). This is just a fancy name for mercenaries who are governed by certain rules and regulations, such as only to engage with the enemy if fired upon first. With the aid of huge government contracts, and wealthy owners, companies such as Blackwater have managed to hoover up many of the world's elite soldiers, provide them with the very best equipment, then offer them to the highest bidder. Such companies are of course very attractive to governments in this day and age, who can now access the best soldiers in the world, for a price.

Moreover, while they don't have to employ these elite soldiers for any long period of time, thus cutting down on costs, they can also engage them on 'unofficial missions'. If anything should go wrong, they can plead ignorance. As was the case in Iraq, governments also don't have to include any PSC deaths in their official casualty lists, making going to war apparently more palatable to their citizens. As such, PSCs can often do a government's dirty work for them, with the public none the wiser.

However, there have been occasions when this has backfired, most notably in Nisour Square, Baghdad, when in 2007 Blackwater soldiers allegedly opened fire on Iraqi civilians, killing seventeen and injuring twenty. Law suits regarding this incident are still ongoing but it was an affair that caused the US government immense embarrassment and proved that employing PSCs might not always be the best solution. Indeed, there are also questions to be asked about certain individuals having such elite private armies at their disposal.

Such has been the advancement in technology in recent years that hand-to-hand combat, or clandestine missions, often don't need boots on the ground at all. Rather than highly trained pilots dropping bombs on enemy countries, as the Luftwaffe did in the Battle of Britain, unmanned drones now do so, controlled from the safety of an air-conditioned office. For instance, US drones have frequently bombed the Middle East in recent years, their pilots controlling them from a facility in the Nevada desert. Although the drones might be shot down, the pilots are now safe and sound, almost reducing warfare to a kind of computer game.

While the likes of Subotai and the Mongol Kheshig had to go on incredible reconnaissance missions to source information on a target, much of this work can now be done by computer hackers. With almost everything kept on computer, governments all over the world engage in cyberwarfare, frequently trying to access each other's secret files, or infect them, without the risk of sending men behind enemy lines. With this, they can also target electricity grids, water networks, financial systems and even hospitals, potentially bringing a country to its knees without a shot being fired.

Spies are of course still in operation but a large part of their role no longer requires face-to-face contact, as most of what they are looking for is online. Intelligence agencies have become particularly adept at sourcing what they need, using programs such as PRISM, which as Edward Snowden, a former agent of the National Security Agency (NSA), revealed monitors everything on the internet. Thus the US is able to obtain virtually any information it desires if it is kept online. Moreover, the job of spying is far easier for agencies than it used to be. With most targets now storing confidential information on their computers or phones, these can be easily hacked, not only allowing the withdrawal of any information but also the tracking of the user. After all, it was by tracking the mobile phone of bin Laden's courier that the CIA were able to pinpoint the terrorist's Abbottabad compound.

It has also been shown that government agencies are able to listen to conversations in homes by hacking into the microphones on items such as televisions. In 2017, Wikileaks claimed that the CIA had used a virus known as 'Weeping Angel' to access Samsung televisions for this very purpose. In 2018, *Wired* magazine also proved that smart speakers, such as Amazon Echo, which are known to store all conversations they record, can also be easily hacked by spy agencies if they so wished. The days of sneaking into a target's home or office and planting bugging devices appear to be over. It seems we now unwittingly do the job for the spy agencies, and pay for the privilege.

In this day and age, many individuals also seem quite happy to put what was once private all over their social media profiles, making it especially easy for agencies to gather information. In 2013, *Vice* magazine revealed that the defence firm Raytheon had developed a computer program called 'Riot', which collated social media 'check-ins' to track a target's movement over Google Maps. It could also collate social networking activity to see who a target was friends with, and what they spent their time doing.

An algorithm would then add all of this information together to predict the behaviour of the target.

Rather than overthrowing governments through assassinating leaders, as the Praetorians had once done, all that seems to be required these days is a social media account. It is alleged that Russia influenced the 2016 US presidential election in favour of Donald Trump by utilising thousands of Twitter bots, and phony Facebook pages, to spread pro-Trump propaganda. In her book *Cyberwar: How Russian Hackers and Trolls Helped Elect a President – What We Don't, Can't, and Do Know*, Kathleen Hall Jamieson states that material generated by the Kremlin had reached 126 million American Facebook users. Moreover, during the election, Wikileaks also published emails that had been hacked from the Democrat servers and were said to be prejudicial to Trump's rival, Hillary Clinton. Many believe that these two things in tandem helped swing the election Donald Trump's way, and, in some people's minds, Russia's way.

Bots spreading propaganda continues to be a major issue on social media. In October 2017 alone, Twitter published more than 10 million tweets by around 4,600 Russian and Iranian-linked propaganda accounts. A 2018 investigation by the *Daily Telegraph* revealed that hundreds of Russian workers man these accounts from a modern four-storey office block in St Petersburg, as part of an ongoing disinformation campaign designed to fray the fabric of western society. It seems social media trolls are the new elite special forces in some circles. In 2018, Facebook tried to address this issue by deleting 583 million fake accounts. This huge number only serves to emphasise the problem we face.

Even as I write this, an AP investigation has apparently revealed that Russian spies have been generating photographs of fake faces to gather information on social media sites, such as LinkedIn. The users' profile pictures are actually eerily realistic faces that have been generated by a computer. It is alleged that one of those who accepted a friend request from a fake Russian account like this was economist Paul Winfree, a former adviser to Donald Trump, who is currently being considered for a seat on the Federal Reserve. China has also been accused of conducting 'mass scale' spying on LinkedIn, with agents sending friend requests to thousands of targets.

While the likes of the Brandenburgers had to go behind enemy lines on hazardous missions, pretending to be the enemy, the rise of artificial intelligence and 'deepfakes' might also render such missions obsolete. This software can perfectly replicate an individual's looks and voice, which can then be used for purposes of impersonation. In May 2019, a video of Donald Trump appeared, apparently offering advice to the people of Belgium on the issue of climate change. The video caused outrage, until it was revealed that the speech was a high-tech forgery. With it being almost impossible to tell the difference between the deepfake and the real thing, this technology could have huge consequences, as we may never know whether news is real or fake.

Even if future conflicts do require elite soldiers to put their boots on the ground, there is no guarantee they will be human. In 2018, a Ministry of Defence report titled 'The Future Starts Today' stated: 'While it is envisaged that humans will continue to be central to the decision-making process, conflicts fought increasingly by robots or autonomous systems could change the very nature of warfare, as there will be less emphasis on emotions, passion and chance.'

Governments have long held ambitions to build robot militaries but until recently it has been the stuff of science fiction. However, in 2018 it was announced that the Pentagon was spending $1 billion on robots to complement combat troops. Indeed, some experts believe that by 2025 the US military will have more combat robots than human soldiers. Russia has also claimed that it hopes to introduce robots to its armed forces as early as this year. It appears that robot soldiers will be with us sooner rather than later.

However, if elite human soldiers are still required, it might be in space rather than on earth. In February 2019, President Trump signed a directive ordering the Pentagon to establish a Space Force as the sixth branch of the United States military, to go along with the US Army, Navy, Air Force, Marine Corps and Coast Guard. Its main goal is to secure and extend American dominance in space. This is truly the next frontier for elite special forces.

Whatever the future holds, I am certain that, in some capacity, the military will always require a specially trained team of elite human beings, who, with all of their guts and perseverance, are capable of staring danger in the face and pulling off the impossible. For, if men like my father can transfer from horse to tank, then I'm sure that today's elite forces can also make the jump, going on to change the world and thrilling us with their escapades in the process.

BIBLIOGRAPHY

THE ROYAL SCOTS GREYS

Royal Scots Greys (Men at Arms) by C. Grant (1972)
Swifter Than Eagles by A. Sprot (1998)
Those Terrible Grey Horses: An Illustrated History of The Royal Scots Dragoon Guards by S. Wood (2015)

THE IMMORTALS

From Cyrus to Alexander, A History of the Persian Empire by P. Briant (2002)
Herodotus: The Histories translated by A. de Selincourt (1954)
History of the Persian Empire by T. A. Olmstead (1948)
Immortal: A Military History of Iran and Its Armed Forces by S. R. Ward (2014)
Shadows of the Desert: Ancient Persia at War by K. Farrokh (2007)
The Achaemenid Persian Army by D. Head (1992)
The Persian Army 560–330 BC by N. Sekunda (1992)
Xerxes' Invasion of Greece by C. Hignett (1963)

THE SPARTANS

A History of Sparta: 950–192 BC by W. G. Forrest (1980)

Elite Military Formations in War and Peace by I. Hamish and K. Neilson (1996)

Military Theory and Practice in the Age of Xenophan by J. K. Anderson (1970)

Thermopylae 480 BC: *Last Stand of the 300* by N. Fields (2007)

Thermopylae: The Battle that Changed the World by P. Cartledge (2006)

The Spartan Army by J. F. Lazenby (1985)

The Spartans: The World of the Warrior Heroes of Ancient Greece, from Utopia to Crisis and Collapse by P. Cartledge (2003)

THE SACRED BAND OF THEBES

An Army of Lovers: The Sacred Band of Thebes by L. Compton (1994)

Military Theory and Practice in the Age of Xenophan by J. K. Anderson (1970)

Sacred Band of Thebes by C. Hilbert (2012)

The Defence of Greece by J. F. Lazenby (1993)

The Rise and Fall of the Sacred Band of Thebes by G. A. Hauser (2011)

The Theban Hegemony: 371–362 BC by J. Buckler (1980)

ALEXANDER THE GREAT AND THE SOGDIAN ROCK

A History of Macedonia II, 550–336 BC by N. G. L. Hammond and G. T. Griffith (1979)

Alexander the Great and his Time by A. Savill (1998)

Arrian: Anabasis of Alexander translated P. A. Brunt (1976)

By the Spear: Philip II, Alexander the Great and the Rise and Fall of the Macedonian Empire by I. Worthington (2016)

Conquest and Empire: The Reign of Alexander the Great by
    A. B. Bosworth (1988)
The Anabasis of Alexander translated by A. de Selincourt (1958)

THE ROMAN PRAETORIAN GUARD

*The Praetorian Guard: A Concise History of Rome's Elite
    Special Forces* by S. Bingham (2011)
The Praetorian Guard by B. Rankov (1994)
The Death of Caligula by T. P. Wiseman (2013)
The Twelve Caesars by Suetonius, translated by R.
    Graves (1957)
*The Praetorian Guard: A History of Rome's Elite Special
    Forces* by S. J. Bingham (2012)
Praetorian: The Rise and Fall of Rome's Imperial Bodyguard
    by G. de la Bédoyère (2018)

THE VARANGIAN GUARD

Harald Hardrada and the Vikings by P. F. Speed (1992)
Harald Hardrada: The Warrior's Way by J. Marsden (2007)
King Harald's Saga by S. Sturluson (1966)
The Varangian Guard, 988–1453 by R. D'Amato (2010)
*The Varangians of Byzantium: An Aspect of Byzantine Military
    History* by B. S. Benedikz (2007)
The Viking Road to Byzantium by E. Davidson (1976)

THE KNIGHTS TEMPLAR AND HOSPITALLERS

Hospitallers: The History of the Order of St John by J.
    Riley-Smith (1999)
Knight Hospitaller (1) 1100–1306 by D. Nicolle (2001)
*Knights of Jerusalem: The Crusading Order of Hospitallers
    1100–1565* by D. Nicolle (2008)
*Saladin and the Saracens: Armies of the Middle East 1100–
    1300* by D. Nicolle (1986)

Saladin: Hero of Islam by G. Hindley (2010)

The Cross and the Crescent by M. Billings (1987)

The Knight Hospitaller: A Military History of the Knights of St John by J. C. Carr (2016)

The New Knighthood: A History of the Order of the Temple by M. Barber (1994)

The Siege of Acre, 1189–1191: Saladin, Richard the Lionheart and the Battle that Decided the Third Crusade by J. D. Hasler (2018)

The Templars History and Myth: From Solomon's Temple to the Freemasons, a Guide to Templar History, Culture and Locations by M. Haag (2008)

The Templars: The Rise and Fall of God's Holy Warriors by D. Jones (2018)

The Third Crusade, 1191: Richard the Lionheart, Saladin and the Struggle for Jerusalem by D. Nicolle (2005)

Warriors of God: Richard the Lionheart and Saladin in the Third Crusade by J. Reston (2002)

THE ASSASSINS

Alamut and Lamasar: Two Medieval Ismaili Strongholds in Iran, an Archaeological Study by V. Ivanov (1960)

A Short History of the Ismailis: Traditions of a Muslim Community by F. Daftary (1998)

Eagle's Nest: Ismaili Castles in Iran and Syria by P. Willey (2005)

Hasan-i-Sabbah and the Assassins by L. Lockhart (1930)

The Assassins by E. Burman (1987)

The Assassins: A Radical Sect in Islam by L. Bernard (2003)

The Old Man of the Mountain by C. Nowell (1947)

The Secret Order of the Assassins: The Struggle of the Early Nizari Ismailis Against the Islamic World by M. Hodgson (2005)

THE MAMLUKS

From Saladin to the Mongols, The Ayyubids of Damascus, 1193–1260 by S. Humphries (1977)
The Mamluks, 1250–1517 by D. Nicolle (1993)
The Mamluks in Egyptian Politics and Society by T. Philipp and U. Haarmann (1998)

THE MONGOL KHESHIG

Genghis Khan's Greatest General: Subotai the Valiant by R. A. Gabriel (2004)
Mongol Imperialism by T. T. Allsen (1987)
The Mongol Art of War by T. May (2007)
The Mongols by D. Morgan (2007)
The Secret History of the Mongols: A Mongolian Epic Chronicle of the Thirteen Century by I. de Rachewiltz (2006)

THE OTTOMAN JANISSARIES

1453: The Fall of Constantinople by S. Runciman (2012)
1453: The Holy War for Constantinople and the Clash of Islam and the West by R. Crowley (2005)
Memoirs of a Janissary by K. Mihailovic (1975)
Ottoman Warfare, 1500–1700 by R. Murphey (1999)
The Janissaries by G. Goodwin (2006)
The Janissaries by D. Nicolle (1995)
The Siege of Constantinople by J. M. Jones (1972)

THE LANDSKNECHTS

Landsknecht Soldier, 1486–1560 by J. Richards (2002)
Pavia 1525: The Climax of the Italian Wars by A. Konstam (1996)
The Landsknechts by D. Miller (1976)
The Landsknechts: German Militiamen from Late XV and XVI Century by L. S. Cristini (2016)

THE NINJA

Hattori Hanzo: The Devil Ninja by A. Cummins (2010)
*Iga and Koka Ninja Skills: The Secret Shinobi Scrolls
    of Chikamatsu Shigenori* by A. Cummins and Y.
    Minami (2013)
*More Secrets of the Ninja: Their Training, Tools and
    Techniques* by H. Kuroi (2009)
Ninja: 1,000 Years of the Shadow Warriors by J. Man (2012)
Ninja: The Invisible Assassins by A. Adams (1970)
The Maker of Modern Japan: The Life of Tokugawa Ieyasu by
    A. L. Sadler (1978)
*The Secret Traditions of the Shinobi: Hattori Hanzo's Shinobi
    Hiden and Other Ninja Scrolls* edited and translated by A.
    Cummins and Y. Minami (2012)

CROMWELL'S NEW MODEL ARMY

New Model Army, 1645–60 by S. Asquith (1981)
The English Civil War by P. Young and M. Roffe (1973)
Cromwell: Our Chief of Men by A. Fraser (1973)
Cromwell's War Machine: The New Model Army 1645–1660
    by K. Roberts (2005)
Oliver Cromwell and the Rule of the Puritans in England by C.
    Firth (1900)
*Soldiers of Parliament: The Creation and Formation of the
    New Model Army During the English Civil War* by C.
    Firth (2015)
Naseby 1645: The Triumph of the New Model Army by M. M.
    Evans (2007)

THE DUTCH MARINE CORPS

1666: Plague, War and Hellfire by R. Rideal (2016)
*A Distant Storm: The Four Days' Battle of 1666, The Greatest
    Sea Fight of the Age of Sail* by F. L. Fox (1996)

Neptune and the Netherlands: State, Economy and War at Sea in the Renaissance by L. Sicking (2004)

The Dutch on the Medway by C. Macfarlane (1897)

The Dutch on the Medway by P. G. Rogers (1970)

The Anglo-Dutch Naval Wars, 1652–1674 by R. Hainsworth and C. Churchers (1998)

The Anglo-Dutch Wars of the 17th Century by C. R. Boxer (1974)

The Anglo-Dutch Wars of the Seventeenth Century by J. R. Jones (1996)

THE BRITISH LIGHT INFANTRY

British Light Infantry in the Eighteenth Century by J. F. C. Fuller (1925)

General Craufurd and His Light Division by A. Craufurd (1906)

How England Saved Europe: The Story of the Great War, The War in the Peninsula by W. H. Fitchett (1900)

Rifles at Waterloo by R. Cooper and G. Caldwell (1995)

The Peninsular War, 1807–1814: A Concise Military History by M. Glover (1974)

The Waterloo Campaign: June 1815 by A. A. Nofi (1998)

Wellington's Peninsula Regiments: The Light Infantry by M. Chappell (2004)

THE IRON BRIGADE

Giants in Their Tall Black Hats: Essays on the Iron Brigade by A. T. Nolan and S. E. Vipond (1998)

The Iron Brigade by J. Selby (1971)

The Iron Brigade: A Military History by A. T. Nolan (1961)

Those Damned Black Hats! The Iron Brigade in the Gettysburg Campaign by L. J. Herdegen (2008)

THE STORMTROOPERS

Stormtroop Tactics: Innovation in the German Army, 1914–1918 by B. I. Gudmundsson (1985)
Storm Troops: Austro-Hungarian Assault Units and Commandos in the First World War by C. Ortner (2006)
The Blitzkrieg Campaigns: Germany's 'Lightning War' Strategy in Action by J. Delaney (1996)
Storm of Steel by E. Junger (1920)

THE RAF AND THE BATTLE OF BRITAIN

Fighter: The True Story of the Battle of Britain by L. Deighton (1977)
Spitfire: A Very British Love Story by J. Nichol (2018)
The Battle of Britain by K. Moore (2010)
The Battle of Britain: The Greatest Air Battle of World War II by R. Hough and D. Richards (2005)
The Narrow Margin: The Battle of Britain and the Rise of Air Power 1930–1940 by D. Dempster and D. Wood (1961)

THE COMMANDOS

Commando by C. Terrill (2008)
Into the Jaws of Death: The True Story of the Legendary Raid on Saint-Nazaire by R. Lyman (2013)
St Nazaire 1942: The Great Commando Raid by K. Ford (2001)
The Raiders: Army Commandos 1940–1946 by R. Neillands (1989)
The Royal Marine Commandos: The Inside Story of a Force for the Future by J. Parker (2007)
The Royal Marines: From Sea Soldiers to a Special Force by J. Thompson (2004)

HITLER'S BRANDENBURGERS

Behind Soviet Lines: Hitler's Brandenburgers Capture the Maikop Oilfields 1942 by D. Higgins (2014)

Brandenburg Division: Commandos of the Reich by E. Lefevre (1999)

German Special Forces of World War II by G. Williamson (2009)

Hitler's Brandenburgers: The Third Reich's Elite Special Forces by L. Paterson (2018)

Kommando: German Special Forces of World War Two by J. Lucas (2014)

THE PARATROOPERS

Airborne: World War II Paratroopers in Combat by J. Guard (2007)

Arnhem, Jumping the Rhine, 1944 and 1945: The Greatest Airborne Battle in History by L. Clark (2009)

Paratrooper! The Saga of Parachute and Glider Combat Troops During World War II by G. M. Devlin (1979)

The Paras: The Birth of the British Airborne Forces from Churchill's Raiders to 1st Parachute Brigade by W. F. Buckingham (2008)

The Paras: The Inside Story of Britain's Toughest Regiment by J. Parker (2000)

THE SAS

Heroes of the SAS: True Stories of the British Army's Elite Special Forces Regiment by B. Davies (2007)

The Originals: The Secret History of the Birth of the SAS in Their Own Words by G. Stevens (2005)

The SAS at Close Quarters by S. Crawford (1993)

The SAS in Action by P. Macdonald (1990)

The SAS: The Official History by P. Warner (1988)

Who Dares Wins: The SAS and the Iranian Embassy Siege by P. Winner and G. Fremont-Barnes (2009)

THE GREEN BERETS

12 Strong: The Declassified True Story of the Horse Soldiers by D. Stanton (2009)

Green Berets at War: US Special Forces in Southeast Asia, 1956–1975 by S. L. Stanton (1999)

Inside the Green Berets: The First Thirty Years by C. M. Simpson III (1985)

Masters of Chaos: The Secret History of the Special Forces by L. Robinson (2004)

United States Special Operations Forces by C. Lamb and D. Tucker (2007)

THE US NAVY SEALS

No Easy Day: The Only First Hand Account of the Navy Seal Mission that Killed Osama bin Laden by M. Owen (2013)

The Operator: The Seal Team Operative and the Mission that Changed the World by R. O'Neill (2017)

US Navy Seals by M. Bahmanyar (2005)

US Special Operations Forces in Action: The Challenge of Unconventional Warfare by T. Adams (1998)

INDEX

anti-aircraft commands, 230,
   285, 288, 313
anti-Nazis, 259–61
anti-tank guns, 321
anti-tank positions, 268, 281, 287
Antioch, 96, 133
AP, 340
Apis calf, 17
Apollo, 63
apple counterbalances, 12
Apries (pharaoh's daughter), 14
Apschetousk, 269
Aqua Claudia, 62
aqueducts, 62
arabaci, 137
Arabia, 11
archers/archery, 13, 28, 49, 51,
   87–8, 127–8, 135–6, 144
Archias of Thebes, 31–2, 46
Arctic Ocean, 48
Arethusa, HMS, 282, 287, 289, 290
Armavir, 267
Armed Forces High Command
   (Germany), *see* OKW
Armenia, 11, 73, 87
Army Air Corps (Britain), 278
Army Group South (Germany), 264
Army Special Forces (US), 310
Arnold, Hugh, 254
Arnórsson (skáld; poet), 76
arquebusiers, 148, 149, 152, 156
Arras, Battle of, 217
Arrian of Nicomedia, 49
Arsuf, Battle of, 93–5, 133
artillery, 134, 136, 152, 209, 215,
   217, 219, 221–4, 230, 269, 277
Ascalon, 90, 96
Asia Minor, 71, 76, 81, 140
Askari people, 259
assassination, 8, 47, 56, 57, 64–5,
   101–9 *passim*, 125, 162, 164–5,
   270, 295, 340
Assassins, 98–109, 159, 162
assault cannons, 219
Assyria, 115
Astyages, King, 9, 10
Athenian democracy, 30
Athens, 18, 23, 30–1, 36–7,
   39, 43, 46
Attic helmets, 57

Auchinleck, Gen Sir Claude, 293–4
Augustus, Emperor, 57–8
Australia, 234
Austria, 196, 220, 222, 259, 265
autonomous systems, 341
axes, 52, 70, 79, 136
Aybak, Izz al-Dīn, 107, 131
Ayyubid dynasty, 87, 88, 93–5
Azerbaijan, 11, 117
Azov Sea, 122
Azuchi Castle, 163

Babylon, 9, 11, 12, 14
Bachait bin Shemtot bin Samra, 35
Bactria, 48, 49
Badger State, 213
Baghdad, 103, 131, 133, 338
Bagram airbase, 333, 336
Baha ad-Din, 92
Bahriyya Regiment, 128
Báibars (Mamluk
   commander), 126–33
Baird, Gen. David, 197
Baku, 269
Baldwin de Boulogne, 82
Balkans, 71, 217, 263
balloon commands, 230
Baltic Sea, 124
Baltic states, 265, 266
Baluch people, 34, 35
Banbury, 169
Banbury, Sir Henry, 195–6
Bangalore torpedoes, 247, 288
Bank, Col. Aaron, 310
barbarian hordes, 28, 43, 203
barrage ballons, 276
Basil II, 71–2, 73, 76
Bastet (goddess), 15, 16
Bataillon Ebbinghaus, 259
Baxter, Richard, 171–2
BBC World Service, 293, 301
Beaverbrook, Lord, 233–4
Beirut, 90
Bela IV, 124
Belgium, 1, 83, 263, 273, 341
Benavente, 201
Benouville, 280, 284, 290
Bentley Priory, 231, 237
Bergen, 183
B52s, 319